Fieldbus and Networking in Process Automation

Fieldbus and Networking in Process Automation

Second Edition

Sunit Kumar Sen

CRC Press
Taylor & Francis Group
Boca Raton London New York

CRC Press is an imprint of the
Taylor & Francis Group, an **informa** business

Second edition published 2021
by CRC Press
6000 Broken Sound Parkway NW, Suite 300, Boca Raton, FL 33487-2742
and by CRC Press
2 Park Square, Milton Park, Abingdon, Oxon, OX14 4RN

© 2021 Taylor & Francis Group, LLC
First edition published by CRC Press 2014
CRC Press is an imprint of Taylor & Francis Group, LLC

Library of Congress Cataloging-in-Publication Data
A catalog record has been requested for this book

ISBN: 978-0-367-71238-9 (hbk)
ISBN: 978-0-367-71342-3 (pbk)
ISBN: 978-1-003-14994-1 (ebk)

Typeset in Times LT Std
by KnowledgeWorks Global Ltd.

Contents

Preface to the Second Edition

Since the first publication of the book almost seven years ago, a lot of developments have taken place in the field of real time fieldbuses. With a view to keep the readers abreast of the current developments, six such fieldbuses have been included in the current edition. Also, some details from the existing book have been pruned to make it more reader friendly. The author sincerely wishes the book to be well accepted by the discerning readers.

1 Data Communication

1.1 INTRODUCTION

Data communication refers to transfer of information from one place to another. The term *data* means an information which is digital in nature, i.e., binary ones and zeroes. Analog data includes TVs, radios, telephone system, etc. In the vast majority of cases, communication is digital in nature, although at the source and at the final user point it is analog in nature.

A communication system includes a transmitter, a receiver and a link connecting these two. The link can be copper wire, optical fiber, and microwave. The link is parallel in nature for short distances, while it is serial for long ones. Data communications involving data are mostly serial in nature. A simplified data communication model is shown in Figure 1.1.

Input information x is the source of data which may be a computer (data) or a telephone (analog). This gives out a data stream $c(t)$ and it is fed to a transmitter block which acts as a modem whose output is $m(t)$. The modem output passes through the transmission medium. $n(t)$ is the signal received by the receiver modem from which data $d(t)$ is extracted, and this is then stored at the final destination point. For a good transmission system, output information, y, should closely match the input information, x.

1.2 COMPARISON BETWEEN DIGITAL AND ANALOG COMMUNICATION

Analog signals are more susceptible to undesired amplitude, frequency, and phase variations. These quantities are of no concern in digital communication where a received signal is analyzed to determine whether a

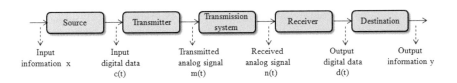

FIGURE 1.1 A simplified data communication model.

signal is above or below a threshold level. In the presence of noise, analog signals are more susceptible to corruption—particularly if magnitude of the signal is low. For such cases, if the impact of noise is not lessened, there are chances that the original signal cannot be retrieved, while a degraded digital pulse can be restored to its original form by employing repeaters along the passage to its final destination. Digital pulses can be stored in memory for later use, which is not possible in analog communication. Transmission rate of digital pulses can very easily be varied to suit different interface conditions. Again, digital pulses can easily be multiplexed and processed. A digital transmission scheme can very easily be evaluated for its performance in terms of bit error rate (BER) which is not so with its analog counterpart. Robustness of digital communication can considerably be enhanced by error detection and correction techniques, data security, data retransmission, flow control—in case there is a possibility of receiver memory being overwhelmed, etc.

On the flip side, a digital system needs precise timing synchronization between transmitter and receiver clocks. An analog signal is at first to be digitized and the corresponding bit pattern is sent. This would entail more bandwidth than sending the analog signal. After that, since an analog signal is to be digitized before transmission and converted back to analog before its eventual use at the receiver—thus conversion error would creep into at either end.

1.3   DATA COMMUNICATION

There are various issues involved in a data communication system which must be addressed for proper data transfer to take place between a source and destination. The following communication tasks are of importance for a multi-source multi-destination communication system: signal generation, synchronization, utilization of transmission facility, error detection and correction, flow control, message formatting, addressing, routing, data recovery, data security, network management, etc.

Signal generation refers to the signal source which must have adequate intensity and proper form such that it can propagate through the transmission medium and is understood by the receiver. Synchronized data transmission scheme is undertaken when considerable amount of data is transmitted so that each separate data packet is received correctly. When a number of devices share the same transmission medium, multiplexing scheme is used to allocate the total transmission capacity amongst the devices. A data transmission scheme *should* include error detection and correction capabilities so that the receiver can detect and correct the actual data even in a hostile transmission environment. In flow control, the receiver lets know the transmitter when to slow down data transmission or else to totally stop transmission. Data formatting refers to a particular form of the data to be transmitted. In a multi-source multi-destination data transmission scheme, addressing of source and destination stations are a must which must be included in the data frame. A data frame may reach the destination via different paths. Routing of data along a specific path is undertaken for proper data delivery to the particular destination. Recovery of data refers to some unforeseen situation in which data transmission is interrupted. The scheme involves resuming the activity at the point at which interruption took place or else to initiate data transmission afresh. Data security involves encryption of data at the transmitting end side such that no third party can have access to the data being sent. At the receiver, the received data is decrypted to get back the original data. The overall communication facility is managed by network management which takes care of present status of the facility, report about failures, configuring the system, etc.

1.3.1 MAIN CHARACTERISTICS

Efficacy of a data transmission system is a function of four characteristics which are *data delivery*, *timeliness*, *accuracy,* and *jitter.* Data delivery involves delivering data at the correct destination only. Timeliness is a quality of a transmission system which ensures timely delivery of data. In case of audio and video transmission, timeliness is the essence of transmission. After transmission, data must be received accurately and correctly at the receiver end, otherwise the received data would become useless. If in a system, data packets arrive at the destination with different delays, then the system would suffer from jitter. Real-time applications, such as teleconferencing, require an upper bound on jitter. The larger the delay variations allowed in a system, there would be corresponding delay in real-time data delivery resulting in greater size of delay buffers at the receivers.

1.4   DATA TYPES

Data can have various forms like numbers, text, audio, video, or images. A number is represented by its equivalent binary. Some important number systems are: decimal or base 10, binary or base 2, hexadecimal or base 16, and IP addresses or base 256.

Text is represented by different bit patterns, with each bit pattern (also known as set or code) representing a particular text symbol. The current coding system uses 32 bits to represent a bit pattern, known as Unicode.

The first 127 characters of the Unicode are used by ASCII (American Standard Code for Information Interchange) code.

Audio and video refer to recording and broadcasting of sound and picture, respectively. While audio is always continuous in nature, video can either be continuous or discrete.

Images are composed of pixels or picture elements. Each pixel is represented by a small dot. An image is divided into a number of pixels. The greater the number of pixels, the better is the clarity of the image. Now, each pixel is represented by a bit pattern. The size and the value of the bit pattern is a function of image quality. Colored image requires a higher value of the bit pattern compared to a black and white image.

1.5   DATA TRANSFER CHARACTERISTICS

Data is converted into digital signal by means of a process called *line coding*. Data or data element is the smallest entity that represents a piece of information. It is represented by bit. Data rate is the number of data elements that can be sent in a second. The unit is bits per second or bps. It could be kbps (kilo bits per second) or Mbps (mega bits per second).

When a bit is converted into Manchester coded form, then in each bit period, there is a transition halfway through the bit period of the coded data. The bit is called the data element while the smallest element of the coded data is called the signal element. The number of signal elements that can be sent in a second is called *signal rate* or *pulse rate* or *modulation rate* or *baud rate*. Thus, signal rate refers to the speed of signal element after encoding and modulation on data is performed.

It should be clearly understood that in communication, it is the data which need to be sent, but it is the signal which actually travels through the link. Increasing the speed of data transmission means more data rate—this increases the throughput. If signal transmission rate can be decreased, it would result in lesser bandwidth requirement of the channel.

Communication speed on the transmission link is limited by bandwidth of the link. It refers to the maximum rate at which signal changes can be handled before attenuation degrades the quality of the signal at the receiving end so much so that it cannot be retrieved faithfully. The difference between the maximum and minimum frequencies contained in a signal is its bandwidth. The *absolute* bandwidth is the difference between the maximum and minimum frequency contained in the spectrum of the signal, while the *effective* bandwidth refers to the difference between the maximum and minimum frequency in the spectrum in which most spectral energy is contained. Bandwidth can either be expressed in Hz or bps. Bandwidth in Hz of a channel is the range of frequencies it can pass without degradation of the signal, while band-width in bps is the number of bits per second that a channel can pass through it.

Latency or time delay is another very important characteristic in message transmission via a transmission link. Latency refers to the time taken by an entire message to reach the destination via various links. There are several components in latency which are: *propagation time, transmission time, queuing time,* and *processing time.* Thus, latency is the sum of all these delays taken together. Less the delay in transmission of a message, the better the system is. Propagation time is the time needed by a single bit to reach the destination from the source. It is measured by dividing the distance by the speed of the medium through which propagation is taking place. Transmission time is the time between the first bit leaving the sender and the last bit of the message reaching the destination. It is defined as the message size divided by the bandwidth. Queuing time is a variable one and depends on the load or traffic through which the data/message has to pass. This is akin to a car taking more time covering a distance during day time when traffic is heavy but takes much less time during morning when traffic is expectedly less. Message from the source to destination has to pass through different nodes. The nodes themselves are to cater to traffic from other sources which pass through them. Thus, depending on the en route traffic, there is always a variable delay for a message to reach the destination.

1.6   DATA FLOW METHODS

Data flow between two communicating devices are done by using simplex (also called *receive only, one way only or transmit only)*, half duplex (also called *two-way alternate* or *either way)* and full duplex (also called *two-way simultaneous, duplex* or *both way)* and full/full duplex modes. These schemes are shown in Figure 1.2(a-d), respectively.

In the simplex mode, communication is always unidirectional—from transmitter to the receiver. In half-duplex mode, two-way data transmission is possible—but not at the same time. In full duplex mode, transmission in both directions can take place between any two stations simultaneously. In full/full duplex mode, simultaneous transmission in both directions is possible between many stations.

1.7   TRANSMISSION MODES

Data transmission between a transmitter and a receiver can take place in a number of ways. It can be parallel or serial. Serial data transmission

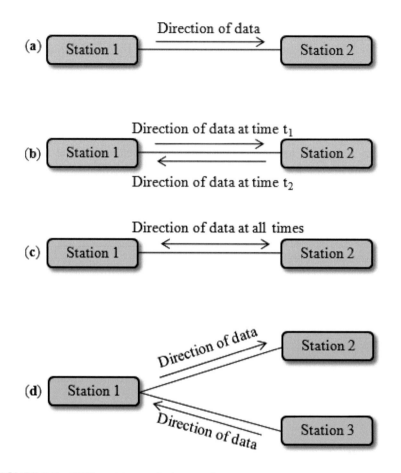

FIGURE 1.2 Different transmission modes.

scheme is again subdivided into asynchronous, synchronous, and isochronous. The different modes of transmission are shown in Figure 1.3.

1.7.1 PARALLEL

A message consisting of n bits requires a single clock to reach the receiver in parallel transmission scheme. Thus, in essence, parallel transmission is much faster than serial transmission, but the cost of laying n wires would entail extra cost. Hence, parallel transmission is undertaken only if the distance between the transmitter and receiver is not considerable.

1.7.2 SERIAL

Serial transmission involves transmission of bits of a message in serial manner—one after the other. When the distance between the transmitter

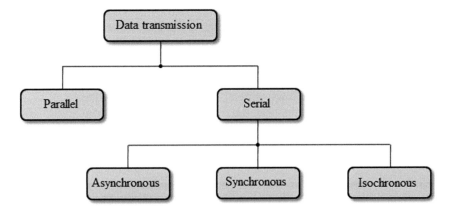

FIGURE 1.3 Different data transfer schemes.

and the receiver is considerable, it makes sense to transmit information serially because parallel transmission would entail considerable cost and laying of wires may sometimes become difficult.

1.7.3 ASYNCHRONOUS

Although termed asynchronous communication, it is basically a synchronous type of communication where synchronization is maintained for each character—each character consisting of 5–8 bits. The receiver resynchronizes at the beginning of each character frame. Synchronization basically means agreeing or coinciding exactly in time scale. Asynchronous communication is called *start-stop* type transmission because framing of each character is between a *start* and *stop* bit(s). A character starts with a start bit (which is always a logic 0), followed by the bits of the data, the parity bit and finally one, one-and-a-half or two stop bits. The stop bit(s) is/are always 1 or high state. Synchronization is achieved at the receiver on receiving the high-to-low transition of the start pulse. The stop bit provides a minimum guard band or buffer period between two characters. An asynchronous frame format is shown in Figure 1.4.

An idle line (i.e., no transmission) is always a stream of 1's. When a character is being sent down the line, it starts with a start bit of status 0,

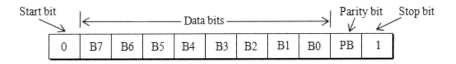

FIGURE 1.4 Asynchronous data frame format.

followed by data bits and finally a stop bit of value 1. Thus, the start bit is identified by the receiver from a high-to-low transition. Assuming a second character follows immediately the first, then again there would be a high-to-low transition at the receiver (due to stop bit of the first character and the start bit of the second character). If no second character is sent, the line continues in the idle line high state after completion of stop bit.

For reliable communication, be it asynchronous or synchronous, a pre-agreed set of rules have to be obeyed by both the transmitter and the receiver, called *protocol*. The major elements comprising a protocol are *syntax* (refers to format and signal levels), *semantic* (refers to control information for proper coordination and error handling), and *timing* (refers to speed matching and sequencing). In the present case of asynchronous communication, the protocols are: clock speed, character frame length, signal level, number of start and stop bits, and type of parity bit (odd or even).

A separate clock line is normally not taken from the transmitter to receiver in asynchronous transmission. However, the receiver clock is designed to be as close to the transmitter clock value as possible. If these two clocks differ somewhat, a *clock slippage* may occur. There can be under slipping or over slipping. The former occurs if transmitter clock is slower than the receiver clock, while if transmitter clock is faster than the receiver clock, over slipping would occur. Thus, for the case of over slipping, the received data is being sampled at a rate slower than the rate at which the data bits are received from the transmitter. In this case, as the samples are being analyzed and stored in the receiver memory, a time will be reached at which a data bit will be completely skipped.

Asynchronous transmission is undertaken when data is sporadic or intermittent in nature and also volume of data to be handled is not huge. This mode is simple and cheap but its overhead is somewhat quite high—about two to three bits for each eight data bits. This comes to about 20–25 percent of overhead bits and hence it affects throughput (it is the number of actual data bits sent in unit time) considerably.

1.7.4 SYNCHRONOUS

In synchronous transmission, considerable data is transported from transmitter to receiver at a fast rate in predefined frames. While transmission is character-by-character type in asynchronous transmission, it is frame-by-frame type in synchronous transmission. In this transmission scheme, it is best to insert clock information in the data signal itself. One such example is Manchester coding. Thus, irrespective of data signal pattern, there would always be a level transition in each period. At the receiving end side, this level transition is utilized to generate the clock signal which would be totally in synchronism with the received data. Thus, even if there

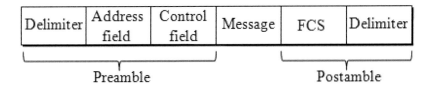

FCS: Frame check sequence

FIGURE 1.5   Synchronous data frame format.

is change in data rate during transmission, no loss of synchronism would occur which is realized by employing phase locked loop (PLL). A synchronous frame format is shown in Figure 1.5. The frame contains several fields. The first field is a delimiter field, also called a synchronous pattern. This pattern alerts the receiver that bits of the message are about to come down the line. This pattern establishes time synchronism with the transmitter clock.

In case of binary synchronous communication (BSC), two synchronous characters are sent, one after the other, to avoid misinterpreting any random data byte to be a synchronous character.

Synchronous transmission becomes more and more efficient as data volume to be transferred increases. Efficiency is the ratio of information bits to total transmitted bits. In this case, the percentage of overhead bits is very low—in the range of 1 percent or less.

1.7.5   ISOCHRONOUS

In synchronous transmission, frames travel down the line in fixed time slots. Here, frames are synchronized for transmission. It fails if there is an uneven delay between frames, which occurs in real-time audio and video. TV transmissions involve sending 30 images per second. TV viewing must also have to be at the same rate. Isochronous transmission guarantees such transmissions such that images arrive at the same rate for the purpose of viewing.

1.8   USE OF MODEMS

The term modem stands for modulator and demodulator. At the transmitting end, it acts as a *signal modulator* whose output is a digitally modulated analog signal, while at the receiving end, it acts as a *signal demodulator* which demodulates the received signal into data. Modems used for data communication purpose is called data communication modem, dataset, data phone, or simply modem.

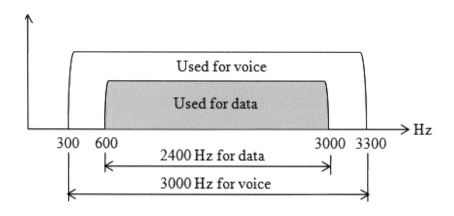

FIGURE 1.6   Bandwidth of voice and data signals.

Normal telephone lines, used for carrying voice signals, operate in the range of 300–3300 Hz, giving a 3000 Hz bandwidth. Voice signals can accept some amount of interference and distortion to the extent that its intelligibility is not lost. Data requires higher integrity for data retrieval and hence the two edges are not used. Bandwidth for data communication is 2400 Hz with lower and upper limits at 600 and 3000 Hz, respectively. This is shown in Figure 1.6. Thus, voice band data modems are used to carry data. In this, bits modulate the carrier, producing digitally modulated analog signals that modulate data into analog form for transmission through the telephone channel using a bandwidth of 2400 Hz of the total telephone line bandwidth.

1.9   POWER SPECTRAL DENSITY

A time limited signal has an infinite bandwidth. However, most of the power of such signals is concentrated within some finite band. In this context, power spectral density is a very important term which shows the power content of a signal as a function of frequency. Effective bandwidth of a signal is the frequency band in which most of the power of the signal is concentrated.

If a channel has high frequency response between two frequencies f_1 and f_2, then the power spectral density of the code which is chosen for transmission should have high power content in this band to avoid signal distortion.

1.10   TRANSMISSION IMPAIRMENTS

Transmission media, which are not ideal or perfect, cause transmission impairments. This means that the signal that is sent is not the one that is received. For analog signals, the impairments cause degradation in signal quality of the received signal while for digital signals, the effect is an error in bits that is received. The most significant impairments are: *attenuation*, *distortion*, *limited bandwidth,* and *noise.*

Attenuation causes a loss of energy of the received signal, i.e., the strength of the signal falls off with distance as it travels down the medium. This loss is due to resistance of the medium. To overcome this, amplifiers are used for analog signals, while for digital signals, repeaters are used.

First, signal received at the receiver must have strength sufficiently higher than noise to eliminate the possibility of any error. Second, attenuation is an increasing function of frequency. Equalizers are employed to overcome this problem across a defined band of frequencies. A second method is to employ amplifiers which amplify high frequencies more than the lower ones.

Distortion changes the form or shape of a signal. A composite signal has different frequencies which travel at different speeds through the medium. Thus, phase relationships that the different frequencies had at the transmitting end won't be maintained at the receiving end. It thus would give rise to delay distortion. It may result in some frequency components of one bit spilling into the next one, resulting in inter symbol interference (ISI). This limits the maximum transmission frequency.

Leaving aside other impairments, larger the bandwidth of the medium, the more closer the received signal will be to the transmitted one.

The last major cause of impairment is noise. There are different types of noise like induced noise, thermal noise, crosstalk, intermodulation noise, and impulse noise. Induced noise comes from operations of industrial motors and electrical appliances. Thermal noise occurs due to random motion of electrons in a wire and is a function of temperature. Crosstalk refers to the effect that a wire causes on a neighboring wire. It is an unwanted coupling between two signal paths. Intermodulation noise may arise when two or more signals of different frequencies share the same transmission medium. It may result in an additional signal of frequency which is the sum of two existing frequencies. Impulse noise generally comes from lightning or power lines. It has considerable energy but exists for a very short time.

1.11   DATA RATE AND BANDWIDTH RELATIONSHIP

Data communication channel is the most important facility in the whole communication process. The greater is the bandwidth of a channel, higher is the cost of the facility. Thus, a given bandwidth must be used as efficiently as possible. For a given level of noise, the maximum data rate is determined by the bandwidth of the channel. A channel has a limited bandwidth which depends on the physical properties of the medium. Data rate through a channel is determined by: available bandwidth, number of levels present in the signal, and also the amount of noise present.

Nyquist bit rate for a noiseless channel and Shannon capacity for a noisy channel are the two yardsticks for determining the maximum data rate through a channel. The maximum theoretical data rate, in bps, is given by: $2 \times B \times \log_2 L$, where B is the bandwidth and L is the number of levels in the signaling element. Thus, for a given bandwidth, if the number of levels is increased, the corresponding bit rate would increase. But data retrieval at the receiver would be difficult in presence of many levels. As per Shannon, the channel capacity, in presence of noise, is given by: $B \times \log_2(1 + SNR)$. SNR is the signal-to-noise ratio at the receiver. This formula does not contain the number of levels present in the signal element. Thus, channel capacity increases with either bandwidth or signal strength. But with increasing signal strength, nonlinearities in the system also increases as also the intermodulation noise. Again, since noise is assumed to be white, the greater the bandwidth, more noise would get into the system. This effectively decreases the SNR.

1.12 MULTIPLEXING

Bandwidth is one of the most precious resources in communication and its judicious use is seemingly the main challenge to the communication engineers. If a low bandwidth (narrow bandwidth) signal occupies a link whose bandwidth is high, then the link's resources are woefully utilized. Multiplexing is an effective way to utilize the available bandwidth.

1.12.1 INTRODUCTION

Multiplexing is the transmission of multiple signals simultaneously over a single link. Although they share the same medium for transfer of information, they do not necessarily occur at the same time or occupy identical bandwidth. Metallic wires, coaxial cables, satellite microwave, optical fiber, etc. may act as the transmission medium. At the receiver end, demultiplexing is done to retrieve the original signals. Figure 1.7 shows the basic principle of operation of a multiplexer–demultiplexer (MUX–DEMUX) system. They are connected by a single link through which N channels

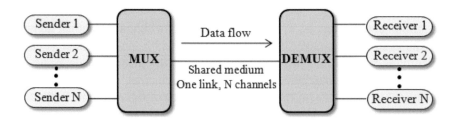

FIGURE 1.7   Schematic of a MUX-DEMUX system.

transmit their information. A multiplexer combines the input signals into a single stream (many-to-one) while a demultiplexer (one-to-many) separates out the signals into individual ones.

1.12.2 TYPES

Three basic multiplexing schemes are there. These are: frequency division multiplexing (FDM), wavelength division multiplexing (WDM), and time division multiplexing (TDM). Out of these, FDM and WDM techniques are used for analog signals, while TDM is used for digital signals. Apart from the above, another multiplexing technique, called space division multiplexing (SDM) is sometimes used. In this, individual cables are allocated for individual signals.

1.12.3 FDM

FDM is a technique in which the available bandwidth in a communication link is divided into a series of non-overlapping frequency sub-bands, each of which carries the modulated version of the original signal. Cable television uses a single cable to transmit many channels for viewing using FDM technique. Other uses of FDM are handling multiple telephone calls through high-capacity trunk lines, communication satellites that transmit different channel data for both up linking and down linking purposes, etc.

In FDM, a guard band is maintained between two adjacent channels which effectively guarantees non-interference between the communicating channels. Figure 1.8 shows six channels with the carrier frequency of the 1st channel fixed at 200 KHz and 6th channel at 1300 KHz with a guard band of 20 KHz between any two adjacent channels.

1.12.4 WDM

WDM is, in a sense, identical to FDM. Optical signals of different wavelengths (i.e., of different colors) are used and sent via a single optical cable. Optical signals have very high frequencies, unlike FDM. Multiplexing ensures that very high bandwidth associated with optical fibers is effectively utilized. A prism is used to bend a beam of light which depends upon

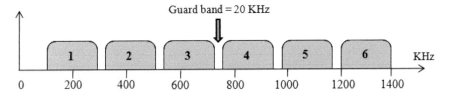

FIGURE 1.8   Different carrier frequencies with guard band in between.

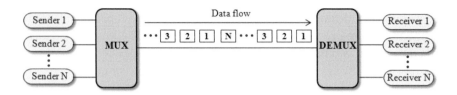

FIGURE 1.9   A TDM MUX-DEMUX system.

the angle of incidence and also the frequency. SONET uses WDM technology in which more than one optical cable is used for MUX-DEMUX purpose. Of late dense WDM (DWDM) is used which uses many channels with very less spacing between them thereby enhancing efficiency even more.

1.12.5   TDM

TDM is a digital communication process which allows several signals from different sources to time share the resources of a link. In TDM, the whole bandwidth of the link is utilized at any instant of time by a single signal while the total bandwidth is always shared by the communicating signals in FDM. In TDM, all signals have their own precise clocks to send data which needs proper synchronization. The different channels have their own scheduling for data transfer. The receiver can extract the channel signals by proper clocking and synchronization with the individual channels. The concept of time division multiplexing is shown in Figure 1.9.

1.12.5.1   Synchronous TDM

In synchronous TDM, data from different sources are divided into fixed time slots, in which a slot may contain a single bit, a byte of data or a predefined amount of data. As shown in Figure 1.10, data from the first source

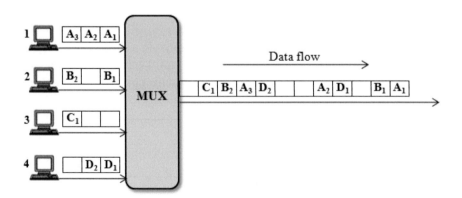

FIGURE 1.10   A synchronous TDM system.

is sent in the first time slot, followed by data from the second source in the second time slot. This is continued until data from the last source is sent. Then the system repeats itself. Thus, in the first four time slots, data A_1 from source 1, data B_1 from source 2, no data from source 3, and data D_1 from source 4 are fed into the multiplexer. In this sequence, the next set of data from the four sources are sent, following the same logic. If some source does not have any data to be sent at any given instant of time, because of pre-allocation, that time is simply wasted. Statistical TDM addresses this problem by skipping slot allocation for a source if it does not have any data in that particular time slot.

1.12.5.2   Statistical TDM

To overcome the shortcomings of synchronous TDM, statistical TDM utilizes only those slots which have data, skipping those slots which do not have data at the times they are to send their data on the channel. Referring to Figure 1.11, one slot for source 2, two slots for source 3, and one slot for source 4 are skipped because they are empty. The scheme must have an identifier to indicate and identify the particular receiver, since the traditional multiplexing scheme is not used here for the sake of improved channel utilization.

1.12.6   VARIABLE DATA RATE

So far the discussions on TDM centered around the assumption that all sources involved in data transmission have the same data rate—which is not the case always. There are different approaches to overcome such data rate variations. These are: multilevel multiplexing, multislot multiplexing, and pulse stuffing technique, which are now discussed in the following subsections.

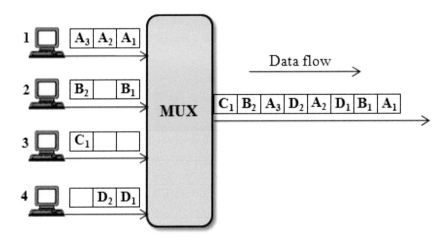

FIGURE 1.11   A statistical TDM system.

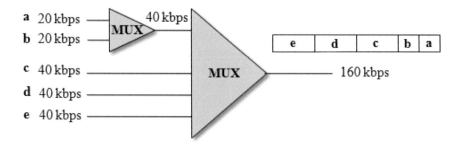

FIGURE 1.12   A multilevel multiplexing scheme. (Courtesy: B. A. Forouzan. Data Communications and Networking, 4th Edition, Special Indian Edition, TMH Companies Inc., New Delhi, India, p. 174, 2006.)

1.12.7   MULTILEVEL MULTIPLEXING

It is employed when most of the lines have higher data rates and fewer ones have slower data rates with data rates of the latter an integral multiple of the former. The slower ones are multiplexed so that the data rate of the multiplexed output is equal to the other ones. Figure 1.12 shows a two-level multiplexing scheme with the two 20 kbps lines combined by the first multiplexer. The second multiplexer effectively combines four input lines. The output frame is a 160 kbps line. It should be noted that demultiplexing at two levels are needed to get back the original five lines—two of 20 kbps and three of 40 kbps.

1.12.8   MULTISLOT MULTIPLEXING

It is employed when most of the lines have slower data rates and fewer ones have faster data rates with the data rates of the former an integral multiple of the latter. Figure 1.13 shows the 50 kbps line divided into two 25 kbps lines so that the reduced rate is equal to the rest three.

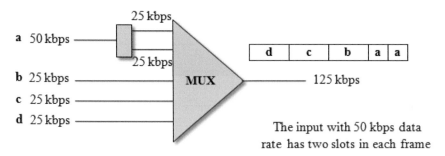

FIGURE 1.13   A multislot multiplexing scheme. (Courtesy: B. A. Forouzan. Data Communications and Networking, 4th Edition, Special Indian Edition, TMH Companies Inc., New Delhi, India, p. 174, 2006.)

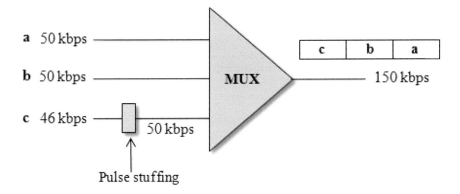

FIGURE 1.14 Pulse stuffing scheme. (Courtesy: B. A. Forouzan. Data Communications and Networking, 4th Edition, Special Indian Edition, TMH Companies Inc., New Delhi, India, p. 175, 2006.)

The 50 kbps line is allocated two time slots in the output data frame of 125 kbps rate.

1.12.9 PULSE STUFFING MULTIPLEXING

The above two schemes fail when data rates of the sources are not an integral multiple of each other. In this scheme, extra bits are added to the slower data rate sources so that ultimately their data rates become equal to the faster ones. This scheme is also called *bit padding* or *bit stuffing* and is shown in Figure 1.14.

1.13 SPREAD SPECTRUM

Spread spectrum technique is used to spread the spectrum of the original signal to camouflage it from possible eavesdropper or being subjected to jamming by some malicious intruder. It was first developed for military or intelligence applications. These secrecy requirements far outweigh the greater bandwidth requirements in this case because of spread of the signal over a larger bandwidth. It is designed for wireless applications like LANs and WANs.

In spread spectrum, an intended signal is converted into a wideband, noise like signal. Such a signal is hard to detect, intercept, and demodulate. This spread-out transmitted spectrum has a power level which may be less than the noise power. Because of this, an unauthorized listener cannot sense anything and the coded message remains invisible. The energy versus frequency characteristics of a spread spectrum signal and the original baseband (BB) signal are shown in Figure 1.15.

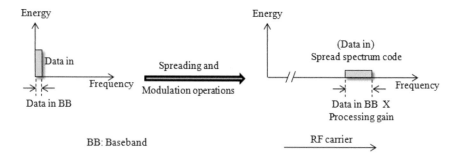

FIGURE 1.15   Energy versus characteristic of an information signal and its spread spectrum signal.

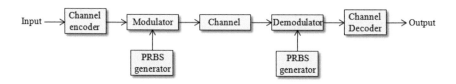

FIGURE 1.16   General scheme of a spread spectrum communication technique.

1.13.1 INTRODUCTION

The general scheme of a spread spectrum communication scheme is shown in Figure 1.16. It includes encoder/decoder, modulator/demodulator, and a PRBS generator. The encoder receives the input and produces a relatively narrow bandwidth analog signal. This signal is further modulated by the modulator with the help of a PRBS generator. The result is a spreading of the bandwidth considerably which is fed into the communication channel. On the receiving side, the received signal is demodulated by the same PRBS generator in a synchronous manner. The channel decoder finally decodes the signal to retrieve the data to get back the original signal.

Two techniques are employed to spread the spectrum: frequency hopping spread spectrum (FHSS) and direct sequence spread spectrum (DSSS).

1.13.2 FREQUENCY HOPPING SPREAD SPECTRUM (FHSS)

In this scheme, FHSS modulated signals hop from one frequency to another in the radio frequency range. The number of hopping frequencies equals the number of channels which are to be multiplexed. At any given instant, a signal modulates only one carrier frequency—the one chosen from the frequency or channel table. The channel usage pattern is called hopping sequence.

Non-return to zero (NRZ) data is initially modulated into either frequency shift keying (FSK) or binary phase shift keying (BPSK) form. A PRBS generates a p bit pseudo sequence where each sequence corresponds to a hopping period. Each sequence in the p bit pattern selects a frequency from a table which is fed to a frequency synthesizer. It generates the frequency selected from the table and acts as the carrier frequency which is modulated by the first modulated output. This modulated output is passed through a band pass filter to get the final FHSS.

The receiver hops between the frequencies in synchronism with the transmitter, as per the agreed protocol. If the hopping period is kept very small, the modulated output hops from one frequency to another very fast and thus any would-be eavesdropper would not be able to decipher the modulated signal. Malicious jamming by inserting noise may jam the signal for one hopping period, but not the whole period. Thus, only one channel may get jammed but not the other ones.

1.13.3   Direct Sequence Spread Spectrum (DSSS)

Like FHSS, this scheme also enhances the bandwidth of the original signal, but in a different manner. Here, each bit in the original message or signal is represented by multiple bits in the transmitted signal—also called *chipping code (or spreading code)*. It spreads the signal in a wider frequency band in direct proportion to the number of bits used in the code. In WLAN, the chipping code used is 11 bits in length, also called *Barker sequence*.

The input NRZ data is modulated by BPSK code while the PRBS output is modulated by a carrier frequency. After this, these two modulated signals are combined to get the transmitted signal. If the original message signal rate is n and the chip code has N bits, the rate of the spread signal is $n \times N$. Thus, a higher value of N would entail a higher spreading of the transmitted signal (spread signal). Immunity to interference is enhanced if each station uses a different chip code.

1.13.4   Comparison between FHSS and DSSS

Although the basic idea remains the same, they have some characteristic differences which distinguish them.

The carrier of the DSSS always remains the same, but FHSS uses different frequencies at different times. Carrier frequency in the case of FHSS hops around the designed band in a predefined manner. FHSS system does not provide any processing gain. It is the increase in power density when the signal is de-spread. Higher processing gain provides for better SNR at the receiving end side.

Compared to DSSS, a FHSS system is much more difficult to synchronize—in FHSS, both time and frequency need to be synchronized at either end while in DSSS, only the timing of the chips needs synchronization. In FHSS, in order to have initial synchronization possible, it parks at a fixed frequency before initiating communication. In the uneventful case of the jammer frequency equaling the parking frequency, there would be no hopping and communication would not be possible. In another instance when communication is going on, it is very difficult to resynchronize once the receiver loses synchronism. In FHSS, latency time is more because the system needs to search the signal in order to lock on to it, while in case of DSSS, it needs a few bits for the same.

FHSS deals with multipath issue in a better manner than DSSS, because in the former the frequency hops from one frequency to another at the end of each hopping period. FHSS is not as robust as DSSS but simpler to detect.

1.13.5　Advantages of Spread Spectrum

There are several advantages which have led to the wide acceptability of spread spectrum techniques in communications. These are now discussed as follows:

A spread spectrum system, if operates in the ISM band, would be allowed to have more power because of its inherent non-interfering nature. Higher transmit power entails higher operating distance than a traditional analog communication system.

By judiciously choosing pseudo random binary sequences, multiple spread spectrum systems can *coexist* without interfering one another. This thus would enhance the bandwidth utility of the system.

Reduced cross-talk interference is another advantage of this system. This is achieved because spread spectrum technique employs processing gain in the receiver. The effect of cross-talk interference can be eliminated if noise below a threshold is rejected altogether, like in voice communication.

Because of atmospheric reflection and refraction, a receiver may receive a transmitted signal via several paths—multipaths. This is shown in Figure 1.17. Multipath fading occurs because of interference between direct path (D) and reflected path (R). However, since despreading process involves synchronization to direct path signal, hence the reflected signal is rejected.

A spread spectrum system is inherently secure in nature compared to both conventional analog and digital communications. Irrespective of FHSS or DSSS, the transmitted signal is totally random in nature, only to

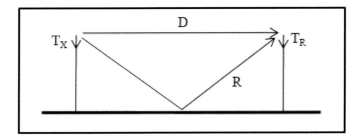

FIGURE 1.17   A signal may reach the receiver in multipaths.

be recognized by the receiver by proper synchronization. Thus, eavesdropping by any malicious intruder is eliminated.

Because of the *pseudo* nature of the code, number of carrier frequencies used as also their values, it is very difficult to decipher the transmitted signal and thus any intruder would be unable to retrieve the original signal.

1.14   DATA CODING

An error- free message received at the receiver is central to any communication scheme. A message, sent from the transmitter, may get corrupted along its passage to the receiver. Many factors play into corrupting message bits—the magnitude of which depends on the type of communication. For audio or video transmissions, some level of error may be tolerated, but in some cases, say monetary transactions must have to have zero level of tolerance. Various schemes are employed to detect and correct transmission errors. Redundant bits, which do not form message bits, are sent along with data bits to achieve the same. Redundancy is realized by various coding schemes that are performed on message bits to ensure error-free message reception. There are several reasons for coding a message before it is sent to the receiver, which are: (a) to simplify transmitter and receiver design (b) to achieve higher transmission efficiency (c) to decrease error and (d) to provide secrecy in transmission, i.e., eavesdropping by a third party is eliminated.

1.14.1   INTRODUCTION

It is very simple to comprehend that error correction is more difficult to realize than simple error detection. Error detection involves the detection of corrupted bit(s). While error correction involves, first, to detect the places of error and subsequently, to correct them. Error correction involves two schemes: forward error correction and retransmission. In the former, the received bits are corrected with the help of redundant bits following some established methodology. Retransmission scheme is straight forward in which retransmission of message is requested when the receiver detects any error in data bits.

1.14.2   CHARACTERISTICS OF A LINE CODE

There are various line codes employed to decrease potential error at the receiver. First, a line code should have the ability to detect and correct errors. Second, the line code should have adequate timing content, i.e., it should be possible to extract clock timing from the coded data which is essential for data retrieval. Third, the code should be transparent in nature. There are situations in which long strings of 1's and 0's are transmitted which form part of data. If it is possible to retrieve such long strings of 1's and 0's from the coded data, then the code is said to be transparent. Fourth, the code should be immune from channel noise and Inter symbol interference (ISI). Finally, frequency response of a coded signal should match the frequency response of the channel through which the coded data is being transmitted.

Some other characteristics that a line code should have are: baseline wandering and DC components. Data, after it is decoded at the receiver, has an average power in its spectrum. This average is called *baseline*. A data element is evaluated for its value after the received signal element power is compared with this average. If a long string of 1's and 0's are there in the data element, there is a possibility of a drift in this baseline value—called *baseline wander*. Such a phenomenon makes determination of data element value very difficult. Thus, a data element should be so coded such that baseline wandering can be avoided. An example of such a code is Manchester coding. When a digital data do not change its status for a few number of bits, there is a possibility of low frequency existence in the spectrum—called DC components. Since AC coupling is used in repeaters, it is expected that the spectrum should be devoid of any DC components.

1.14.3   TYPES

Various line codes are used for sending messages to the receiver to achieve error free reception. Manchester code is a very good example of a transparent code for which there is always a change in signal level at half the bit period. This code is particularly suitable for data having long strings of 1's and 0's. Huffman codes utilize probability of occurrence of different characters and assign shorter codes to characters that are transmitted most often.

Coding schemes can broadly be categorized into block coding and convolution coding. Messages are divided into several blocks—each such block is called a data word. To each such data word, redundant bits are added to get code words. In block coding scheme, each identical data word gives rise to same codeword, i.e., it is a one-to-one coding scheme. Block coding can be employed for both error detection and correction.

Linear block codes are a form of block codes which are employed for both error detection and correction. For such a linear block code, XORing of two valid code words would give rise to another valid code word. A cyclic code is a special type of linear block code in which if a code word is cyclically shifted, it would result in another code word. CRC is a type of cyclic code. If automatic repeat request (ARQ) protocol is used in conjunction with CRC code, then it becomes very effective in reducing Bit error rate (BER) of a message. A CRC-16 code may attain one undetected error in every 10^{14} bits.

2 Networking

2.1 INTRODUCTION

The field of communication has expanded at a very fast pace over the last several decades. Initially only voice communication was in vogue, but availability of microprocessors, microcomputers, and peripheral equipment has drastically changed the whole scenario. Today, a multitude of information, mostly in digital form, is transported all over the globe in a fraction of a second almost without any error. Very high volume of data along with faster, reliable, and secure communication is the need of the hour.

The original information from the source can be in either analog or digital form. If it is analog, it is first to be digitized and then sent. A data communication system involves transmission, reception, and extraction of the original data after processing.

A network is a set of devices (often termed as nodes or stations) interconnected by communication links. Examples of a node can be a computer, a printer, or any device which is capable of sending/receiving data to/from other nodes connected with the network. A data communications network essentially does not have any limit with regard to its capacity. Data communications networks connect a workstation with mainframe computers, ATMs to bank systems, airline and hotel bookings, electronic mail transfers, information highways, media and news centers, etc.

Figure 2.1 shows the structure of a computer communications network. The black circles are switching computers which are connected by high-speed lines. There are also computers connected to different kinds of terminals and some do not have any connections associated with them.

Again, there are three types of terminals: local, remote, and orphan—the names are apparent from their positions in the figure. The local computers operate at moderate bandwidth for data transmission while the high-speed switching computers need higher bandwidth for their very fast nature of operation.

2.2 CHARACTERISTICS

A data communications network must fulfill the following criteria or attributes for an effective and reliable network.

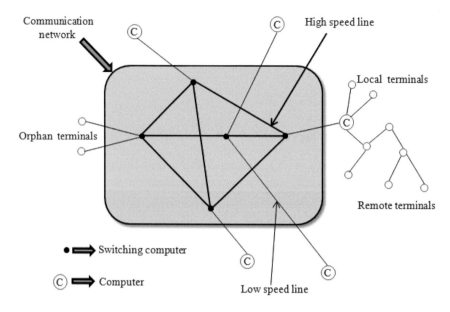

FIGURE 2.1   The structure of a computer communications network.

- **Fastness:** Data transmission from a host must be as fast as possible. Throughput is a measure of fastness. As network traffic increases, throughput decreases and delay increases.
- **Transmit and response times:** They must be kept as low as possible. Transmit time is the time for a message to travel from one node to another, while response time is the time between an enquiry and the corresponding response.
- **Cost:** The cost of establishing, maintaining, and operating the network must be a minimum.
- **Efficiency:** A major part of the network must not remain idle for a long time.
- **Secure:** Any malicious intruder must not have access to the network.
- **Error:** The system must be as error free as possible.

2.3   CONNECTION TYPES

A network consists of two or more nodes connected by a communication link. A link is a path through which information travels from one node to another. Two types of connections are there for transfer of information from node to node: point-to-point and multipoint or multidrop. These are shown in Figures 2.2 and 2.3, respectively.

FIGURE 2.2   A point-to-point connection.

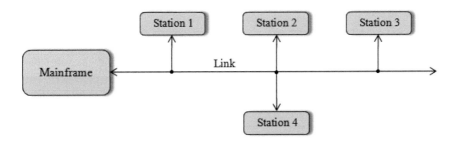

FIGURE 2.3   A multipoint connection.

2.4   DATA COMMUNICATION STANDARDS AND ORGANIZATIONS

Standards are essential for data and telecommunications industry so that equipment manufacturers abide by these agreed standards. This would ensure national and international interoperability between devices from different manufacturers. Standards, guidelines, and frameworks are established by a consortium of organizations, government agencies, manufacturers, and users of equipment in communication industry. The standard organizations in the data, telecommunications and networking industries are: International Standards Organization (ISO), International Telecommunications Union – Telecommunications Sector (ITU-T), Institute of Electrical and Electronics Engineers (IEEE), American National Standards Institute (ANSI), Electronics Industries Association (EIA), Telecommunications Industry Associations (TIA), Internet Architecture Board (IAB), Internet Engineering Task Force (IETF), and Internet Research Task Force (IRTF).

2.5   NETWORK TOPOLOGY

Network topology refers to the manner of interconnecting computers, cables, and components—both physically and logically. Physical connection refers to the actual laying of the network while logical connection refers to how data actually passes from one node (station) to another.

The different topologies used for networking purposes are: mesh, star, bus, ring, and hybrid. All these are examples of multidrop topology.

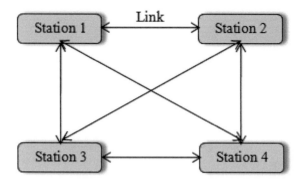

FIGURE 2.4   A mesh network.

2.5.1 MESH

In mesh topology, each station or node is connected, via a point-to-point link, with every other node. To realize the above, each node in an n mesh network must be connected to the rest $(n - 1)$ nodes. Thus, a total of $n(n - 1)$ physical links are needed. However, for a full duplex mode, where communication takes place simultaneously in either direction, a total of $n(n - 1)/2$ duplex mode links are needed. Figure 2.4 shows a mesh network consisting of four stations.

There are several advantages associated with employing a mesh network. First, because of dedicated links, data can move from one link to another irrespective of system load. Second, data security/privacy is ensured. Third, the system is robust in the sense that a fault in a link does not impact other parts of the system. Fourth, faulty node can very easily be detected. Fifth, a traffic diversion can be done if a faulty node is detected.

Because of point-to-point connection, with increase in number of nodes, both cable length and consequently cost increase and huge amount of cabling may lead to space problem.

2.5.2 STAR

Here each station or node is connected to a central computer, called a hub. It acts as a multipoint connector. Any data exchange between two nodes takes place only via the central computer. This is shown in Figure 2.5. A hub has store-and-forward facility—this enables it to handle more than one information at a time.

Some advantages are associated with a star network. First, the system is robust in nature. That is, if a link fails, data exchange between that particular node and the hub fails—without affecting any other node. Fault detection

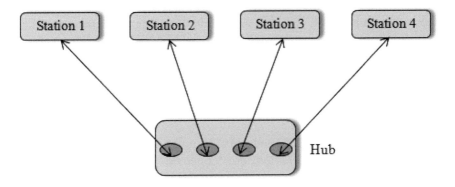

FIGURE 2.5 A star network.

and isolation is very easy. Second, less cabling leads to lesser cost. Third, reconfiguration is very easy in the case of addition/alteration/deletion of any node to the system.

  If the central computer or the hub fails, the star network fails. Again, depending on certain situations, more cable would be needed. Star topology is normally used in local area networks (LANs).

2.5.3 Bus

In this case, there is a long cable—acting as the backbone from which small drop cables (lines) are connected to individual stations or nodes. This connection is made by means of taps. As signal travels along the backbone, it becomes weaker because of presence of taps. This puts a limit to the number of taps that bus technology can support and the distance involved. The scheme is shown in Figure 2.6. Bus topology is sometimes referred to as *linear* or *horizontal* bus. Bus topology normally involves a centrally placed host computer which controls data/information flow to and from other computers.

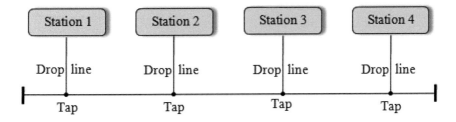

FIGURE 2.6 A bus network.

Bus topology requires less cable length. But fault detection/isolation/re connection is difficult. It is also difficult to add extra nodes to an existing system. There exists also a possibility of signal quality degradation at the taps because of signal reflection. Any fault or break in the trunk or bus cable leads to a total breakdown of the whole system.

2.5.4 RING

Ring topology has a dedicated point-to-point connection between any two successive devices. Stations are connected in series to form a closed loop. It is shown in Figure 2.7. Traffic in the ring is unidirectional. Information, meant for a particular station, is passed along the ring in one direction. This is not accepted by stations along the route for which it is not meant. Instead, they regenerate the signal with the help of the respective repeaters until it reaches the destination station or node. Ring topology, like bus and star topologies, normally has a centrally located host computer which controls data/information flow to other stations in the network.

It is advantageous to use ring topology because it offers easy installation and reconfiguration. Fault isolation is very easy. Whenever any addition/deletion of nodes is required, maximum ring length along with the maximum number of devices used in the ring topology are the only issues to be borne in mind.

Unidirectional traffic in a ring network is a major disadvantage. Any break at any place in the network will bring the entire network down. This can be overcome by employing dual cables (rings) or switches which has the capability of closing off a break.

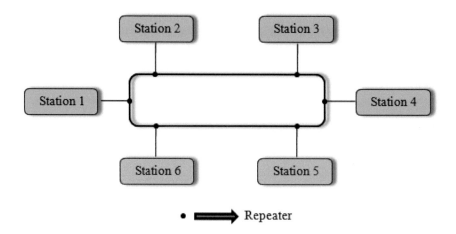

FIGURE 2.7   A ring network.

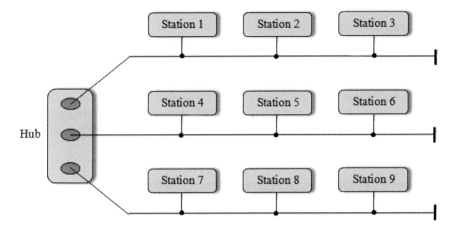

FIGURE 2.8   A hybrid network.

2.5.5   HYBRID

It is normally a mix of the topologies discussed already. A combination of star and bus topology is shown in Figure 2.8. It shows a network in which three lines are star connected to the hub. Stations or nodes are connected to each of these three lines. A hybrid topology combines the advantages of both star and bus topologies.

2.6   NETWORK APPLICATIONS

Networking is all about sharing resources and data between computers over the communication links. The important parameters of a communication network are reliability of transmission, speed, data security, and performance.

A network must be designed with the specific application in mind. A network designed for a specific application would work nicely for that application category, but may not work so if the application profile changes. Table 2.1 shows

TABLE 2.1
Different Networking Applications

Standard office applications	E-mail, file transfer, data storing, printing, etc.
High-end office applications	CAD, software development, video imaging, high volume data transfer, etc.
Manufacturing automation	Process and discrete control, CNC machines, etc.
Mainframe connectivity	PCs, workstations, servers, etc.
Multimedia applications	Live interactive video, etc.

Source: Courtesy: W. Tomasi. Advanced Electronic Communications Systems, 6th Edition, Pearson Inc., New Delhi, India, p. 127, 2004.

the various network applications which have been categorized as per different needs.

Network applications are categorized into connection-oriented and connectionless protocols. Considerable amount of overhead in the form of encryption, error checking, and handshaking slows down the speed of data transmission rate in a major way for connection-oriented protocols while in the case of connectionless protocols, these reliability features are dispensed with thereby increasing data transmission rate in a big way.

2.7   NETWORK COMPONENTS

Various network components help in proper operation of the network. These are: servers, transmission media, shared data, clients, shared printers and other peripherals, network interface cards (NIC), hardware/software resources, local operating system (LOS), and network operating system (NOS).

A server is a computer processor which provides specific services to the network. Users can access network resources by means of a server. Different kinds of servers provide different services like routing server, gateway server, terminal server, web server, printer server, file server, mail servers, communication servers, database servers, etc.

A routing server connects nodes and networks having like architectures. Gateway servers connect nodes and networks having different architectures with the help of protocol conversions. A file server provides a copy of the file from the server to the client (user) who has requested the same.

Transmission media are also called links, channels, or lines. They can be coaxial cables, twisted wire pair, or optical fiber and are used to interconnect computers to the network.

Shared data are used by end users (computers) and are provided by file servers. Some examples are: e-mail, printer access programs, etc.

Computers stationed at either end of the network are the clients/users/customers which take the network facilities for proper end-to-end process data delivery. Clients request for requisite services from the servers and receive the same and execute data transfers.

Each computer on the network has an NIC, whose purpose is to format, process, control, transmit, and receive data to and from the network. At the transmitter, NIC passes the formatted data to the physical layer. At the receiver, NIC receives the formatted data from the physical link, strips the overhead from it, and process the same as per its contents. NICs must have a driver to operate on the network to which it is connected.

LOSs permit computers to access files, CDs, and printers as per the needs. Some examples are MS-DOS, PC-DOS, Windows 95, 98, and 2000.

NOS is a software which allows computers to interact with the network. Thus, it runs on both computers and servers. An NOS provides password authentication, network administration functions, data file and printer access, etc. Some examples of NOS are IBM LAN Server, UNIX, Microsoft Windows NT Server, Novell NetWare, etc.

2.8 CLASSIFICATION OF NETWORKS

Networks are classified by their sizes. It encompasses the geographical area, speed, number of computers, media, and the physical architecture. Initial primary classification includes LANs, metropolitan area networks (MANs), wide area networks (WANs), and global area networks (GANs). Some more primary types of networks are building backbone, campus backbone, and enterprise network. In addition, personal area networks (PANs) and power line area networks (PLANs) are fast catching up in the area of networking. In PANs, data transfer takes place via human body by simply touching them while PLANs use existing AC distribution networks to transmit data from one place to another.

2.9 INTERCONNECTION OF NETWORKS

An Internet is a collection of a few networks connected together which can communicate and share their resources and data traffic. On the other hand, Internet is a public data communication network consisting of hundreds and thousands of networks, used by millions of users. Private individuals, public agencies, governmental organizations, university and research organizations, schools, colleges, airlines, railways, libraries all over the world use the Internet to share data and information. Internet/intranet is executed by browsers such as Microsoft Explorer, Netscape Navigator, etc. Internet came into being only in 1969 at ARPA (Advanced Research Projects Agency). Internet, intranet and World Wide Web (WWW) have together created a virtual explosion in communication arena for data and information transfer at a very high speed with higher and higher traffic density. WWW is a server-based application which allows subscribers to have access to services offered by the Web.

An accurate representation of Internet is very difficult because of its continual change—some networks are added/deleted to/from the Internet always. The Internet, as it is seen today, consists of a complex hierarchical structure. At the lowest rung of this structure lie the LANs connecting the end users. The end users are connected to the Internet via the local Internet service providers (ISPs). Local ISPs are connected to regional ISPs via complex switches. Regional ISPs are in turn connected to national ISPs via switches again. Figure 2.9 shows the hierarchical organization of the Internet in which Figure 2.9(a) shows the details of the structure of a national ISP while Figure 2.9(b) shows the hierarchical structure of interconnections of national ISPs via network access points (NAP).

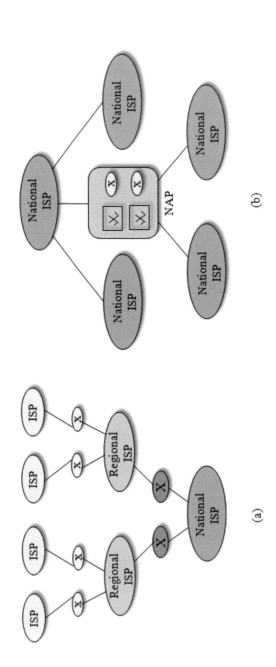

FIGURE 2.9   Hierarchical organization of the Internet. (a) Hierarchical structure of a national ISP. (b) Interconnection of national ISPs. (Courtesy: B. A. Forouzan. Data Communications and Networking, 4th Edition, Special Indian Edition, TMH Companies Inc., New Delhi, India, p. 18, 2006.)

3 Network Models

3.1 INTRODUCTION

A network is all about interconnecting two or more devices together so that data/information can flow either way. When the devices are from the same manufacturer, it is rather easy to interconnect them because they follow the same set of rules, specifications, and guidelines. Communication between two devices from different manufacturers normally face hardware problems as well as software incompatibilities. Such a system is called a *closed* or *proprietary* system, while for an *open* system, communication between two devices from different manufacturers do not face any problem. Such an open system is termed as *interoperable*.

Thus, for such an *open* system, the specifications and guidelines are *open* to all devices connected to the network. The set of protocols underlined in an open system allows any two different systems to communicate with each other.

All networks, be it standard, proprietary, or open, are defined by the *International Standard ISO/IEC 7498-1:1994 Information Technology-Open Systems Interconnection – Basic Reference Model: The Basic Model*. This was first introduced in 1986. This model can be applied to all communication systems—right from PCs to satellite systems.

3.2 A THREE LAYERED MODEL

Before the introduction of seven layer open systems interconnection (OSI) model, a three layer model was conceptualized, involving an application layer, a transport layer, and a network access layer.

Two levels of addressing are needed for an application residing in a computer to reach correctly to another computer. First, applications in a computer must have their own separate addresses, called *service access point* (SAP), which would enable the transport layer to support multiple applications in a computer. Again, each computer connected to the network must have a *unique address*. This enables the network to deliver data at the proper destination computer.

The application layer supports different applications with the help of a software dedicated for these applications. A three layer model involving four computers is shown in Figure 3.1. Each computer in the network has its own software to support application, transport, and network layers.

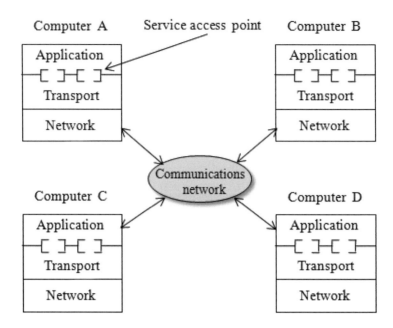

FIGURE 3.1 A three layer communication model. (Courtesy: W. Stallings. Data & Computer Communications, 6 th Edition, Pearson Inc., New Delhi, India, p. 16, 2000.)

The transport layer helps in exchanging data from one computer to another reliably. It ensures delivery of data at the exact destination application. Control information existing at the transport layer ensures proper data delivery.

The network access layer or simply network layer helps in exchanging data between the computer and the network. This layer provides the address of the destination computer to which data is to be ultimately delivered. It is the responsibility of the network to route the data traffic on the network properly so that it reaches its destination. This means that the other two layers are not concerned about the specifics of the control software inherent in network layer. Again the network layer does not know the service access point at which the data is to be finally delivered at the destination computer. Different software are used at this layer depending on the type of the network—circuit switching, packet switching, local area networks (LANs), etc.

Figure 3.2 shows how application data passes through transport and network layers by adding control information at the respective layers (called headers). Protocol data unit (PDU) of a layer is the combination of control information (also called header) of the layer and the total block from its just upper layer. Header information in the transport PDU include: a destination SAP, sequence number, and error detection mechanism. With the help of the destination SAP, the receiver computer can direct the received data at

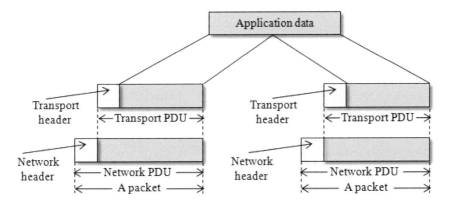

FIGURE 3.2 Transport and network PDUs. (Courtesy: W. Stallings. Data & Computer Communications, 6 th Edition, Pearson Inc., New Delhi, India, p. 17, 2000.)

the correct application file. Sequence number is essential when the transport PDU is sending a series of the same. It helps in rearranging the received data in correct order in case they are received otherwise. Finally, a transport PDU includes a code that helps in detecting whether the received data has been received correctly or not. The receiver can then take action accordingly. Again the network PDU would include a network header that contains the destination computer address and some extra facilities, if needed. The network PDU, with the help of its header, guides the received data to the destination computer. The network header may include some extra facility which would help in prioritizing some message to be transported ahead of others.

3.3   THE OSI MODEL

An OSI reference model provides a common basis which aids in development of system interconnection standards. The model covers all aspects of network communication as envisaged in International Standards Organization (ISO). ISO is an organization and not a model whereas OSI reference model is not a protocol or a set of rules, but it is an overall framework which forms the basis to define protocols. OSI has a layered architecture that facilitates design of network systems allowing communication between all types of computer systems. It consists of seven layers and is shown in Figure 3.3.

The seven layers are divided into three subgroups. Layers 1, 2, and 3 are known as *network support layers*, while layers 5, 6, and 7 are called *user support layers*. The layer lying in between, i.e., layer 4, links the two subgroups.

Application layer provisions user interface using HTTP, FTP. Presentation layer is responsible for presenting data and handles encryption. Session

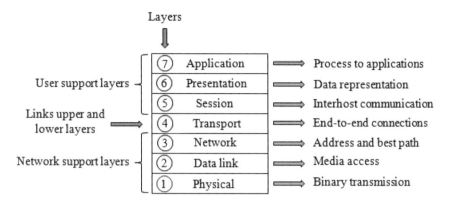

FIGURE 3.3   The seven layer OSI model.

layer keeps different application data separate and is responsible for synchronization. Transport layer provides connection oriented (TCP) and connection less (UDP) end-to-end delivery of data segments and error correction. Network layer provisions logical addressing guiding routers to route data packets to reach destination via the shortest path. Data link layer converts data packets into frames, performs error detection (and not correction) and provides access to media using MAC address (802.2 and HDLC). Physical layer provides bits to the physical medium at specified voltage level (EIA/TIA-232 V.35).

While developing the model, similar types of networking functions were clubbed together and put into a specific layer. This way, different layers were assigned different functions and an architecture developed which is comprehensive and at the same time flexible also. Since the functionalities of each layer is separate and well defined, standards can be developed independently and simultaneously—thus speeding up the process of standardization. Again, since the layers are independent of each other, any change in standard in one layer would not affect the existing software in another layer.

Source data is encapsulated in packet form which starts at the upper layer and moves down the successive layers, adding control information in the form of header at each layer and also trailer at the data link layer. When the packet (with headers added at each layer and trailer) reaches layer 1, i.e., physical layer, it is sent across a physical communication link that passes through the intervening nodes before finally reaching the destination station. This is shown in Figure 3.4. The process of getting back data at the destination node is done by reversing the sequence that was followed in the source node side.

Figure 3.5 shows data exchange between two computers with headers and trailer placed at appropriate places in each layer.

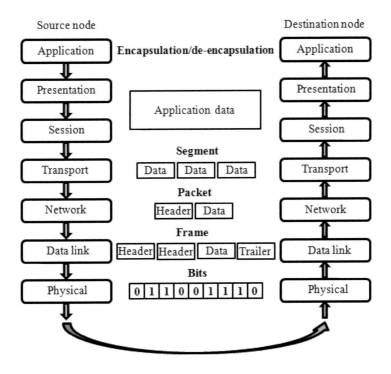

FIGURE 3.4   Data transfer between devices and peer-to-peer processes.

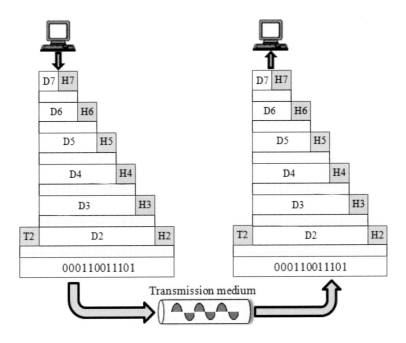

FIGURE 3.5   Data exchange using the OSI model.

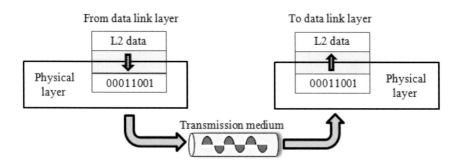

FIGURE 3.6 Operation of physical layer.

3.3.1 PHYSICAL LAYER

Figure 3.6 shows the mechanism of transporting data from the physical layer on to the physical medium. Physical layer receives data from the data link layer. It deals with the physical and electrical specifications of the interface and the medium, as also the functions and procedures that the physical devices and interfaces have to perform for transmission to take place.

Characteristics associated with the physical layer are:

- It transforms bits into signal, i.e., how the 0s and 1s are to be encoded for transmission over the physical medium.
- It defines data rate or transmission rate.
- It is the duty of the physical layer to synchronize the transmitter and receiver clocks.
- It defines the physical topology—i.e., how the devices are connected, viz., mesh or star or ring or hybrid.
- Line configuration, i.e., either point-to-point or multipoint configuration of the devices is the responsibility of the physical layer.
- It is concerned with the mode of transmission, i e., simplex, half-duplex, or full-duplex.
- It defines the characteristics of the interface between the devices and the transmission medium.
- It provides the necessary specifications for different types of hardware like cabling, connectors and transreceivers, network interface cards (NICs), hubs, etc.

3.3.2 DATA LINK LAYER

The data link layer is shown in Figure 3.7. Characteristics associated with data link layer are:

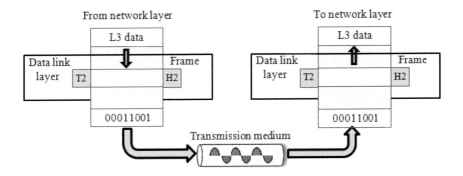

FIGURE 3.7   Operation of data link layer.

- It divides the whole message received from the network layer into smaller manageable data units—termed as *frame*.
- It moves the frames from one node to another node (hop-to-hop).
- It imposes a *flow control* mechanism when data produced by the sender is at a higher rate than the rate at which data can be absorbed by the receiver.
- The layer has an *error control* mechanism by virtue of which it can detect and retransmit damaged or lost frames. It also can recognize duplicate frames. It is achieved by adding a trailer at the end of each frame.

Data link layer is subdivided into an upper sublayer called *logical link control* (LLC) and a lower sublayer called *media access control* (MAC). LLC is responsible for flow and error control. LLC ensures that protocols like IP can function regardless of type of physical technology used. Multipoint access is resolved by MAC, i.e., MAC acts as a mediator. Technologies which are used to achieve the above are: carrier sense multiple access with collision detection (CSMA/CD) for Ethernet and token for token ring systems.

Data link layer adds a header to a frame which needs to be distributed to different systems. Then the header includes sender and receiver addresses. This is known as *physical addressing.*

3.3.3   NETWORK LAYER

The mechanism of data flow through the network layer is shown in Figure 3.8. The responsibilities of network layer include:

Network layer takes the responsibility for the source to destination delivery of a packet across multiple networks.

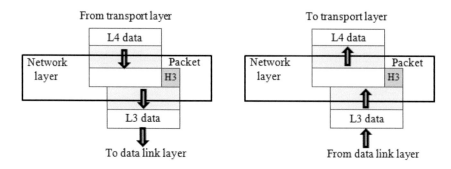

FIGURE 3.8 Operation of network layer.

If a packet, residing in a network, is to be sent to another network, the network layer then adds logical (network) address of the sender and receiver to each packet.

Such addresses are assigned to local devices by a network administrator. This is assigned dynamically by a special server called dynamic host configuration protocol (DHCP).

Several networks are connected by routers and switches to form a large network. The network layer identifies the best path to route the packets to the final destination.

3.3.4 TRANSPORT LAYER

The transport layer is shown in Figure 3.9. Responsibilities carried out by the transport layer are as follows:

- It ensures process-to-process delivery of an entire message.
- While network layer treats each packet independently, the transport layer treats the entire message en masse and sees to it that all the packets are in order.

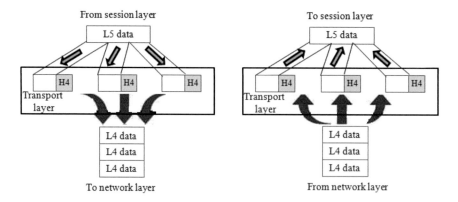

FIGURE 3.9 Operation of transport layer.

- The transport layer can be either connectionless or connection oriented. In connection-oriented transmissions, the receiving device sends an acknowledgement back to the source after a packet is received. This is not so for a connectionless transmission, Thus, while the former is a slower transmission method, the latter is relatively faster.
- A computer may run several processes at the same time. A transport layer header assigns a port address to each such process.
- Transport layer divides a message into segments, with each segment containing a sequence number. The sequence numbers enable the assembly of the message at the receiver. It also identifies and replaces packets lost in transmission.
- Flow control in transport layer is end-to-end rather than a single link.
- Error control in transport layer is process-to-process rather across a single link.
- Transport layer protocols include transmission control protocol (TCP) and user datagram protocol (UDP). The former is connection oriented while the latter is connectionless.

3.3.5 SESSION LAYER

Figure 3.10 shows the operation of the session layer. It acts as the *dialog controller* for the network. Its job includes establishing, maintaining, and synchronizing and finally terminating the interaction amongst the devices which communicate with each other. If a session is broken, it attempts to retrieve the session.

Responsibilities carried out by the session layer include:

Dialog control involves determining which of the two devices are to communicate data between themselves. Data sharing may be simplex, half-duplex, or full-duplex.

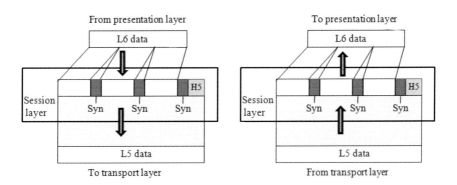

FIGURE 3.10   Operation of session layer.

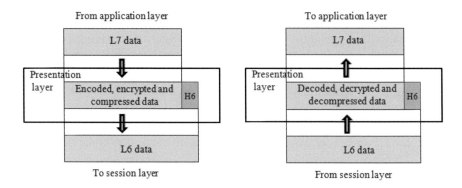

FIGURE 3.11   Operation of presentation layer.

Session layer adds checkpoints, also called synchronization points, to a stream of data. For a huge amount of data, it adds checkpoints in between at predetermined intervals to ensure that data up to each consecutive checkpoints are received and acknowledged properly. The process of adding checkpoints and markers to a stream of data is called *dialog separation*.

3.3.6 PRESENTATION LAYER

Syntax and semantics of a data message are taken care of by the presentation layer. Figure 3.11 shows the operation of a presentation layer. Its responsibilities include:

- Presentation layer ensures that data encoded differently by different computers are interoperable.
- Sensitive information, to be exchanged between sender and receiver, must be kept away from possible eavesdroppers. Data is encrypted in a manner that hides the information from data poachers. Decryption is done to transform the message back to its original form at the receiver.
- Data compression is a method which reduces the bit numbers contained in a data stream, without losing vital information.
- Presentation layer formats include: Text (ASCII, EBCDIC, RTF), Images (JPG, TIF, GIF), Audio (MP3, WAV), Movies (MPEG, AVI, MOV), etc.

3.3.7 APPLICATION LAYER

Figure 3.12 shows the operation of the application layer whose main characteristics are given as follows:

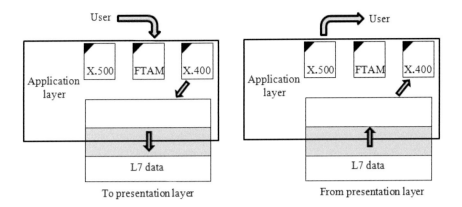

FIGURE 3.12   Operation of application layer.

- It provides user interface and supports various services like e-mail, file transfer, and access to the world wide web.
- It allows users to locate data from a remote location, retrieve the same and use the same at the user's place.
- An user can log into a remote computer and use its resources.

3.4   TCP/IP PROTOCOL SUITE

TCP/IP protocol suite was in vogue prior to OSI model. The original TCP/IP had four layers and obviously did not match the layers of the OSI model. When OSI model was introduced, there was a belief that it would overwhelm TCP/IP commercially, which it never did. Ultimately, during 1990s, it became apparent that TCP/IP would dominate the commercial protocol architecture on which further developments would be based.

3.4.1   INTRODUCTION

TCP/IP was developed by the Department of Defense prior to introduction of the seven layer OSI model. It is the *de facto* global standard for the Internet. The Internet (earlier known as ARPANET) was part of a military project of Advanced Research Projects Agency (ARPA) and the communication model based thereon is known as ARPA model. ARPA was developed in the USA before the OSI model was developed in Europe by the ISO. Whereas the OSI model specifies exactly what function(s) each layer has to perform, TCP/IP comprises several relatively independent protocols that can be combined in several ways.

It is not mandatory to use all the layers in the TCP/IP model, for example, some application-level protocols operate directly on top of IP. TCP/IP does not include the bottom network interface layer, but it depends on the

TABLE 3.1
The Five Layer TCP/IP Reference Model

Layer no.	Layer name	Purpose
5	Application	Specifies communication methodology between different processes/applications residing on different hosts
4	Transport	Provides end-to-end reliable data transfer
3	Network	Routes data between host and destination nodes through one or more networks connected by routers
2	Data link	Concerned with logical interface between an end system and a network
1	Physical	Concerned with physical transmission medium, signal encoding scheme, signal transmission rate, etc.

same for access to the medium. TCP/IP is a hierarchical protocol meaning that each upper layer protocol is always supported by one/more lower level protocols.

3.4.2 PROTOCOL ARCHITECTURE

The five layer TCP/IP reference model is shown in Table 3.1.

The application layer corresponds to the upper three layers of the OSI model, i.e., application, presentation and session layers. TCP residing in the transport layer ensures data delivery to the proper process. Network layer routes data from the host to the destination node via one or more networks, with the help of the IP addresses. The data link layer interfaces an end system with the network while the lowest or physical layer is concerned with signal rate, signal encoding, etc.

Figure 3.13 shows the protocols available at different layers in the TCP/IP protocol suite.

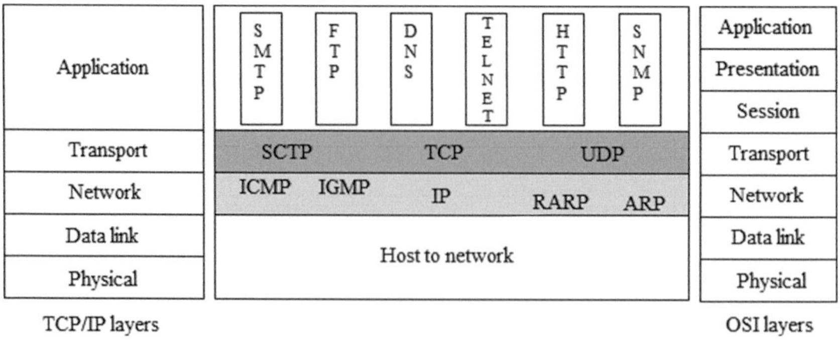

FIGURE 3.13   Protocols in the TCP/IP protocol suite.

At the different layers, different protocols are available. Amongst them, TCP and UDP belong to the transport layer and IP in the network layer are the ones which form the basis for data delivery from one computer stationed at one end of the globe to another computer housed at the other end.

3.4.2.1 TCP

TCP is a *connection oriented transport layer protocol.* Functionalities include reliable data delivery, congestion control, duplicate data suppression, flow control, etc. Most of the user application protocols like FTP and Telnet use TCP. At the transport layer, TCP/IP uses three protocols: TCP, UDP, and stream control transmission protocol (SCTP). Two processes can communicate with each other through the TCP connection with the help of IP datagrams. This is shown in Figure 3.14.

TCP establishes a session between the transmitting and the receiving processes before initiating transmission. There are facilities to check that all packets have been received and arrange for retransmission, in case of packet loss. These involve additional overhead and leads to higher processing time and header size but at the same time makes the system more reliable.

TCP fragments a large chunk of data into smaller segments when necessary, numbers the segments, reassemble the whole message, detects and arrange for retransmission in case of failure, issues acknowledgements for data received, provides socket services for multiple connections to ports on

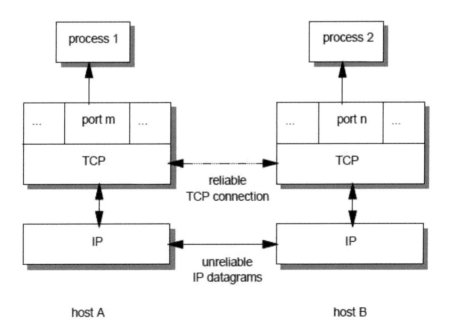

FIGURE 3.14   Connection establishment between two processes by TCP.

0	15	16	31

Source port	Destination port
Length	Checksum
Data	

FIGURE 3.15 UDP header format.

remote hosts. The segmented messages may be received out of order at the receiver, which the TCP reassembles in correct order.

Each and every TCP segment has a header which comprises all the necessary information for proper data delivery and retrieval.

3.4.2.2 UDP

UDP is a connectionless protocol. It does not require any connection establishment prior to data transmission. UDP does not require any sequence number, timers, synchronization parameters, retransmission of data packets, and priority options. Thus, it has less overhead. Its major drawback is that it does not guarantee delivery. UDP is normally used for broadcasting, general network announcements, real time data, etc.

The UDP header is shown in Figure 3.15. It has only four fields.

The source port is an optional one. When it is used, it indicates the port address of the sending process. When not used, a value of zero is inserted for this field. The destination port indicates the process to which the data is to be delivered. The *length* is the length in bytes of the used datagram, including the header. The checksum is an optional 16-bit field, used for validation purposes.

3.4.2.3 IP

Internetworking layer in TCP/IP has some very important protocols which are: Internetworking protocol (IP), Internet control message protocol (ICMP), address resolution protocol (ARP), and DHCP. They together perform datagram addressing, routing, delivery, dynamic address configuration, and resolve between internetwork layer addresses and the network interface layer addresses. IP sends data in packets, called *datagram*. Packets (or datagrams) can travel along different paths (routes) and may arrive out of sequence. IP does not have the facility to rearrange the packets at the receiving end.

IP is an unreliable, connectionless, and best effort packet delivery protocol. Best effort delivery means that packets sent by IP might be lost, may reach out of order, or even may get duplicated. It is the responsibility of the higher layer to address these concerns. Connectionless network protocol is used to minimize the dependence on specific computing centers that uses hierarchical connection oriented networks.

IP addressing is a must to identify a host on the Internet. Thus, each host is assigned an *IP address* or an *Internet address.* A host is recognized by this IP address. A host may be connected to more than one network, called *multihomed*, in which case the host must have a separate address for each network interface.

IP addresses are represented by a 32-bit unsigned binary value and is expressed in a dotted decimal format. Each IP address consists of *a network number* and a *host number.* The network number is administered by one of the three Regional Internet Registries (RIR): American Registry for Internet Numbers (ARIN), Reseaux IP Europeans (RIPE), and Asia Pacific Network Information Centre (APNIC). For example, 128.3.7.8, 128.3 is an IP address, 128.3 represents the network number while 7.8 represents the host number. Sometimes, terms like *network address* or *netID* is used instead of network number while *host address* or *hosteID* is used for host number.

An IP datagram (the basic data packet exchanged between hosts) contains a *source IP address* and a *destination IP address.* For a datagram to be sent to a destination IP address, it must be translated or mapped into a physical address. For example, in LANs, the IP address is translated into physical MAC address by address resolution protocol (ARP).

There are five classes of IP addresses: A, B, C, D, and E—it depends on the number of hosts and network size. Delivery of datagrams using IP addresses can be any of the following types: unicast, broadcast, multicast, or anycast. This is shown in Figure 3.16.

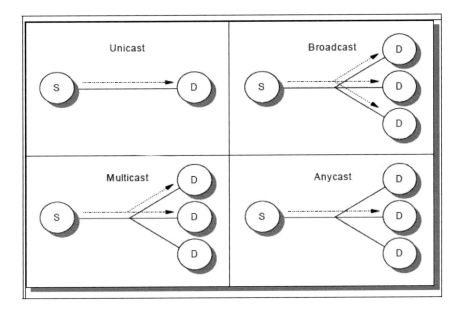

FIGURE 3.16   Different packet delivery modes.

IP version 4 (IPv4) is responsible for delivery of data packets (datagrams) between sending and receiving hosts. Ethernet can handle 1500 bytes while X.25 can handle 576 bytes. Because of this frame size limitation, a message is broken up into fragments, called datagrams. To each datagram is given an IP header, which is then sent from the sending host. The receiving host rebuilds the message from the datagrams received. IPv4 header consists of at least five 32 bits *long words*, i.e., 20 bytes in all. This IP header is appended to the information that it receives from higher level protocols.

3.4.3 OPERATION

TCP/IP protocol suite helps in sending a message from a process associated with a port residing at a host to another process associated with a port at a second host. The receiving host may reside on the same network or on another network. For the latter case, the message has to pass through several routers along its passage to the final destination. It should be borne in mind that IP is implemented in all the end systems and the routers while TCP is implemented only in the end systems.

As already mentioned in Section 3.2, two levels of addressing are needed for a process data in one host to be sent to another process in another host. A local port address is needed which could ensure correct data delivery at the process at the receiving host. Again a network address is needed which would enable the message to be delivered to the receiving host.

Let's say a process data residing at a port belonging to a host is to be delivered to another process having its own port address and belonging to another host. The sending end process hands the message down to TCP. It has instructions to send the same to the second host at the particular port. TCP hands over the message to IP with instructions to deliver the same to the second host. IP is remaining to be totally transparent about the port address of the destination host. All these are managed by control information appended to the message at each layer of the TCP/IP protocol suite.

3.4.4 PDUs IN ARCHITECTURE

Control information, in the form of headers and trailers, are appended to the message at different layers to ensure proper data delivery at the proper destination with utmost reliability. Figure 3.17 shows data encapsulation and PDUs in the TCP/IP architecture. At the TCP layer, control information,

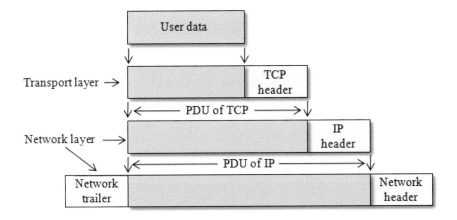

FIGURE 3.17   Data encapsulation and PDUs in TCP/IP architecture.

in the form of TCP header, is appended forming a TCP segment or a PDU of TCP. Control information which are included in the TCP header are: destination port address, sequence number, and checksum. Similarly, at the network layer, an IP header is added, giving rise to PDU of IP. The IP header includes destination network address. It may include some other control information like priority in data delivery, etc.

3.4.5   ADDRESSING

Addressing through the TCP/IP protocol involves sending data from one process to another via the Internet. The addressing involves the following categories: *Physical* or *Link*, *Logical* or *IP*, *Port*, and *Specific*. The addresses refer to specific layers in the TCP/IP model and shown in Figure 3.18.

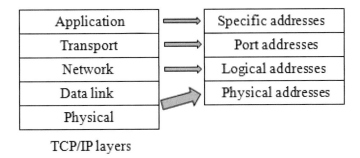

FIGURE 3.18   Addresses and their corresponding layers in the TCP/IP model.

3.4.5.1 Physical

The physical or link address is the lowest level address. It is the address of a station or node specified in its frame by LAN or WAN. Depending on the type of the network, the size and the format of the address vary. Ethernet uses a 6 byte address which is embedded in its NIC.

3.4.5.2 Logical

Logical address corresponds to the network layer in the TCP/IP model. The physical address may vary—depending on the type of the network. Logical address overcomes this difficulty by recognizing a host irrespective of the physical address type. At present, a 32-bit logical address can uniquely recognize a host connected to the Internet. No two IP addresses can be same so that two different hosts can be differentiated and recognized with their logical addresses.

3.4.5.3 Port

A computer may run several processes at the same time. It may communicate with a second computer via a file transfer protocol, message handling services, or TELNET. Thus, these processes residing on a computer must have individual addresses to receive data from other computers simultaneously. This is taken care of by port addresses. A port address is 2 bytes in length. Thus, it is the port address on a computer which helps in exact data/message delivery meant for a particular process once it has reached its destination host.

3.4.5.4 Specific

Specific addresses are user friendly addresses like e-mail address or an URL (Universal Resource Locator). An e-mail address locates a particular recipient in any part of the world while an URL helps in locating some document/writing/information available from the World Wide Web.

4 Networks in Process Automation

4.1 INTRODUCTION

Over the last few years, the processing speed of computers has increased many folds and the associated networking technology has improved beyond comprehension. Initially, it was local networking with only educational institutions and large business conglomerates used to share information amongst themselves. The Internet came in handy to share data/information globally in a reliable, secure, and fast manner. Industrial communication is all about transporting field information to the control room and control a process reliably and effectively. There are several network protocols that include both hardware and software to ensure robust, reliable, and sometimes real-time operations depending on the situational needs. Generically, industrial networks are referred to as fieldbus which includes Sensorbus and Devicebus also—the reference to a particular bus depends on the data size.

There is a growing demand amongst the users of industrial automation systems for an industrial protocol which is vendor independent. Some key factors determine the acceptability of an industrial network protocol which are: an open system, reduction in the cost of wiring and increased information availability from the field devices.

4.2 COMMUNICATION HIERARCHY IN FACTORY AUTOMATION

In an industry—be it a manufacturing or process one, data or information flows from field to the upper layers up to the management/enterprise level and vice versa. For an orderly flow of information and optimize the performance of the process, the whole set up is divided into several hierarchical levels. This is shown in Figure 4.1.

The lowest or the field level comprises sensors, positioners, and actuators. The sensors give an indication about the process condition—temperature pressure, flow, etc. of the process—it can be digital, analog, or hybrid as in HART. The second or the I/O level marshals inputs and outputs together. The sensor output, via the I/O level, is directed to the controller which then generates the appropriate control signal and is fed back to the actuator

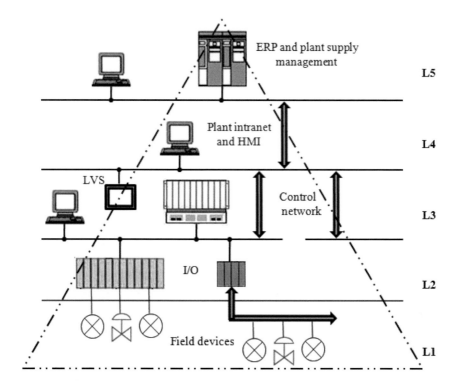

FIGURE 4.1   Hierarchical levels in process industry.

for control of the process. At the control level, normally programmable logic controller (PLC) or distributed control system (DCS) are used which generate the control signals. These signals are sent to the sensor/actuator level where control action is taken—may be opening or closing of valve, starting or stopping of motors, etc. Network engineering tools, plant maintenance, schedule programs, asset management etc. are the jobs of the plant intranet/human machine interface level. Human machine interface ensures that any process variable value from any place in the plant can be displayed on the operator console, can warn the operator in case any process value undershoots/overshoots predefined set limits, can change the device configurations, etc. At the top-most enterprise/supply chain management level, information flows into the office environment for record keeping, billing via service access point, ordering of raw materials, keeping a record of quality and quantity of finished products, etc. and a record of present stock position of different raw materials to keep the production going.

The different levels have to handle different requirements peculiar to that particular level concerned. For instance, the enterprise level has to handle a large volume of data which are not time critical, and is not in constant use either.

The lower three levels have different characteristics like constant use, short data packet length and response time, deterministic communication, etc. Of course there are differences between manufacturing and process automation in these three layers. Whereas in process automation, speed is less and the signal contains more status information, high speed of transmission and short data lengths characterize the manufacturing industries.

Because of characteristic requirements existing at different levels in the hierarchy, it is obvious that no single protocol can address all these. Speed of data transmission, data volume, and security are the major concerns of these levels. Irrespective of whether it is manufacturing or process industry, data must be made available at the HMI and enterprise levels. For efficient and reliable data exchange between the different levels, standardization in the form of ISO-OSI reference model is used to overcome the above issues.

4.3 I/O BUS NETWORKS

The decentralized control and distributed intelligence in the smart field devices have given rise to I/O bus networks. Data and status information are communicated through the I/O bus, for example, the ON/OFF state of a switch, the condition of a machine or process data are sent via the I/O bus network—be it manufacturing or process industry. Figure 4.2 shows

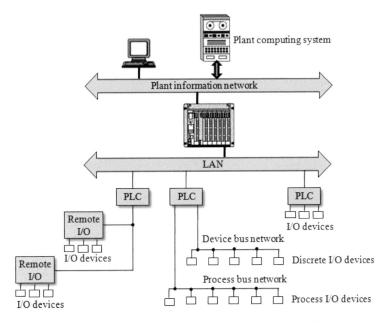

FIGURE 4.2 Interconnections between PLCs, LAN, and I/O bus networks. (Courtesy: Industrial Text and Video Company. I/O bus networks-including Device Net, 1999. Available at www.industrialtext.com)

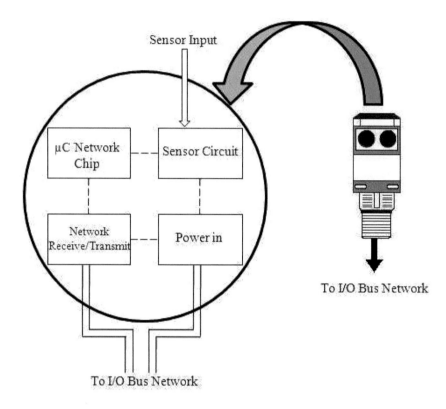

FIGURE 4.3 Intelligent field device used in I/O bus network. (Courtesy: Industrial Text and Video Company. I/O bus networks-including Device Net, 1999. Available at www.industrialtext.com.)

the inter connections between a supervisory PLC, local PLCs, a local area network (LAN), and an I/O bus network.

The process bus network and the device bus network are connected to a PLC. Remote I/Os are connected to other PLCs, as shown. These PLCs are connected to a LAN which in turn is connected to a supervisory PLC and then taken to the other higher layers like HMI and enterprise levels.

Field devices connected to the I/O bus networks must be intelligent in nature and such a field device has, within it, a sensor along with its processing circuitry, a microcontroller, transmitter/receiver network and a power module. This is shown in Figure 4.3.

4.3.1 TYPES

I/O bus networks are divided into two categories: device bus networks and process bus networks. Device bus networks interface with low-level

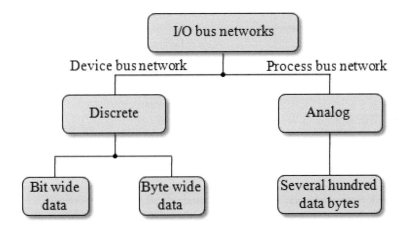

FIGURE 4.4 Classification of I/O bus networks. (Courtesy: Industrial Text and Video Company. I/O bus networks-including Device Net, 1999. Available at www.industrialtext.com.)

information devices (which are usually discrete devices like limit switches, push buttons, etc.), while process bus networks interface with high-level information devices (which are usually smart sensors, control valves, pressure meters, etc.). Device bus networks handle from a few bits to several bytes of data while process bus networks handle considerable amounts of data—a few hundred bytes at a time. Figure 4.4 shows the classification of I/O bus networks.

Process bus networks and device bus networks mostly deal with analog and discrete devices, respectively. Examples of device bus network dealing with analog devices are thermocouples and variable speed drives—devices that transfer a few bytes of information at a time.

Bit wide data bus networks handle up to 1 byte of data at a time while byte wide data bus networks handle between 1 and 50 bytes or more of data at a time. Bit wide bus networks are also called sensor bus networks.

Process bus networks and device bus networks have differing data transmission requirements. Discrete devices transmit only a small amount of data at a time and thus meet the high-speed requirements of such devices. Process bus networks are relatively slow because their data packets contain substantial amount of information which are transmitted at a time.

4.3.2 NETWORK AND PROTOCOL STANDARDS

In the sphere of I/O bus networks, several organizations are working together for establishing network and protocol standards. Organizations like Instrument Society of America (ISA) and European International Electronics Committee (IEC) are involved in developing the standards. In

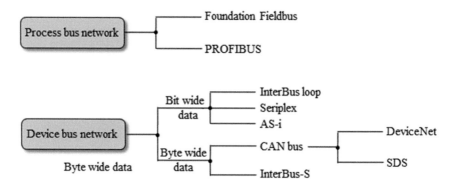

FIGURE 4.5 Network and protocol standards. (Courtesy: Industrial Text and Video Company. I/O bus networks-including Device Net, 1999. Available at www.industrialtext.com.)

the process bus area, two main organizations, namely Fieldbus Foundation (it was established after a merger of Interoperable Systems project (ISP) and the World FIP North American Group) and Profibus (Process Field Bus) Trade Organization are working towards establishing protocol standards. Figure 4.5 shows the different protocol standards under I/O bus network.

In the device bus network, several bus protocol standards exist with Seriplex, AS-i, InterBus Loop under the ambit of bit wide data network and CANBUS, DeviceNet, SDS, InterBus-S under byte wide data network.

4.3.3 Advantages

Advent of digital communication has ushered in widespread acceptability of I/O bus networks in process automation. Digital communication allows more than one fieldbus device to communicate their data over a wire. This is due to addressing capabilities inherent in smart and intelligent devices which provide their process data value, device status information, health etc. Digital transmission has its own advantages because they are less susceptible to electromagnetic interference. Again it is very easy to restore a degraded digital signal to its original value, unlike its analog counterpart.

A second major advantage of digital communication is that such an intelligent field device can pass a digital value directly proportional to the process value—thus eliminating the need to either linearize or scale the process data.

Other established advantages of I/O bus networks are considerable reduction in wiring, the easiness with which plant expansion or upgradation is possible, considerable cost savings, reduced downtime, and very easy fault identification and isolation—if necessary.

4.4   THE OSI REFERENCE MODEL

The OSI reference model has already been discussed in Section 3.3. It is a benchmark on which current and future communication standards can be evaluated. An open system is a set of protocols which would allow two different communication systems to interact with each other regardless of architectural differences. The OSI model is divided into seven different functional layers or levels having clearly different tasks. Each layer has a predefined task. This ensures compatibility amongst the devices using a particular protocol and also between devices using different protocols. Table 4.1 gives a detail of the OSI layers from the perspective of fieldbus.

TABLE 4.1
OSI Model from Fieldbus Perspective

Layer	Function	Task	Standards/Realizations for Fieldbuses
7	Application	Provides the user with specific network commands	DHCP, SNTP, SNMP
6	Presentation	Encodes layer 7 data before forwarding it to layer 5 and decodes layer 5 data before forwarding it to layer 7	Not relevant for fieldbuses
5	Session	Synchronizes communication sessions between two applications	Not relevant for fieldbuses
4	Transport	Prepares data string for transmission and ensures that it is reliably exchanged	TCP, UDP
3	Network	Selects data routes and ensures that the network is not overloaded	IP
2	Data Link	Establishes and maintains connections between two participants	IEEE 802.2, Token passing, IEEE 802.3, CSMA/CD IEC 61158/PROFIBUS-PA : Master/slave, IEC 61158/FF: LAS
1	Physical	Pushes data into the physical medium and takes it off at the destination	EIA RS-232, EIA RS-422, EIA RS-485, IEC 61158-2 100Base T (IEC 800.3u)

Source: B. G. Liptak. Instrument Engineers' Handbook, Process Software Digital Network, 3rd Edition, CRC Press, Boca Raton, FL, p. 433, 2002.

Data or message from the source travel down from layer 7 or application layer, accepting header information from each layer. The physical layer determines speed, transmission medium, etc. from which information packet travels through the medium before being accepted by the receiving medium.

4.5   NETWORKING AT I/O AND FIELD LEVELS

With a bidirectional control reaching up to the level of the field devices in case of fieldbus technology, process control has undergone a sea change in the last two decades. Control and I/Os are now embedded into the field devices. In the field control system (FCS), there are only two levels existing: field level network and host level network. In the current scenario and in the existing plants, 4-20 mA analog communication is still in vogue and would continue to be so for some more time.

The binary field devices like limit switches, solenoids, two position open-close valves, etc. are integrated via the binary I/O cards of PLCs by a point-to-point connection. Within a fieldbus controller, remote I/Os will be there to integrate these signals. For Modbus and AS-i, appropriate interface must be there in the controller for integrating such binary field devices. The 4-20 mA signals are connected through the analog I/O cards in the PLCs in a point-to-point manner. Figure 4.6 shows the connection to the PLC for simple 4-20 mA analog signals along with 4-20 mA hybrid, i.e., HART signals. Legacy field devices are connected to the control network via intelligent I/Os.

In Foundation Fieldbus, field devices are attached to the Foundation Fieldbus H1 segment as shown in Figure 4.7. A linking device connects this

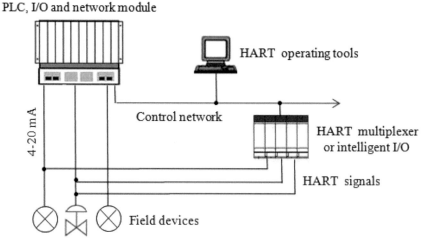

PLC, I/O and network module

4-20 mA

HART operating tools

Control network

HART multiplexer or intelligent I/O

HART signals

Field devices

FIGURE 4.6   Connections of analog and hybrid signals to a PLC.

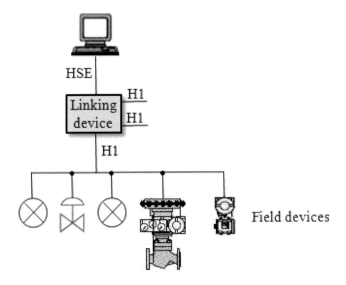

FIGURE 4.7   Connections of foundation fieldbus segment to controller.

bus segment to the high-speed Ethernet (HSE) or H2. The linking device performs only protocol or baud rate conversion. In case of PROFIBUS, devices are connected via a PROFIBUS segment (PROFIBUS PA) which are then taken to the higher speed network (PROFIBUS DP) via a segment coupler or link. The link acts as an interface between the two segments. This is shown in Figure 4.8.

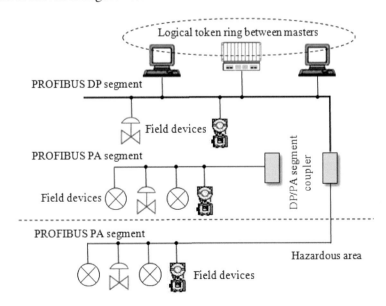

FIGURE 4.8   Connections of PROFIBUS segment to controller.

4.6   NETWORKING AT CONTROL LEVEL

In process automation, the two competing fieldbuses—Fieldbus HSE and PROFIBUS-DP along with ControlNet govern the control level in the automation hierarchy.

Foundation Fieldbus is specially designed for process automation in which the lower speed H1 is connected to higher speed HSE via a linking device. In the case of PROFIBUS, a segment coupler couples the PA network to the higher speed DP. There are some other proprietary protocols which enjoy considerable market share, apart from Interbus which is a manufacturing protocol used in bottling and packaging industries.

4.7   NETWORKING AT ENTERPRISE/MANAGEMENT LEVEL

Ethernet TCP/IP network is used at the enterprise/management level for internetworking and sharing data with the outside world. For this, hubs, switches, routers, etc. are used. Standard data exchange formats are used for inter-company data sharing, e-business applications, etc. For data/ information exchange between management and control levels, quite often PLCs are used to which an Ethernet port is attached for connection to a supervisory station.

5 Fieldbuses

5.1 WHAT IS A FIELDBUS

Fieldbus is a digital two-way multidrop communication link between intelligent field devices. It is a local area network (LAN) dedicated to industrial automation. It replaces centralized control networks with distributed control networks and links the isolated field devices like smart sensors/transducers/actuators/controllers. FOUNDATION Fieldbus H1 and PROFIBUS PA are the two fieldbus technologies used in process control.

In this two way communication, it is possible to read data from the smart sensor and also write data into it. The multidrop communication facility in fieldbus results in enormous cable savings and resultant cost reduction. A fieldbus device must have a fieldbus interface unit for proper communication to take place and shown in Figure 5.1.

Figure 5.2 shows a conventional point-to-point system and its fieldbus counterpart, while a comparison between a 4-20 mA system and a fieldbus system is shown in Table 5.1.

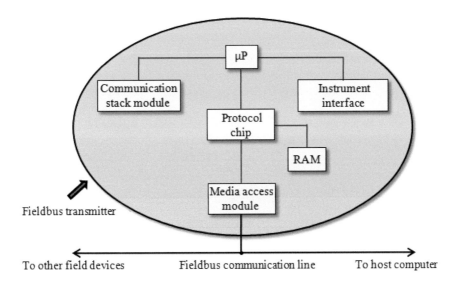

FIGURE 5.1 A fieldbus interface unit in a fieldbus transmitter. (Courtesy: Rosemount Inc. The Basics of Fieldbus. Technical Data Sheet. Chanhassen, MN, 1998, Available at http://www.rosemount.com)

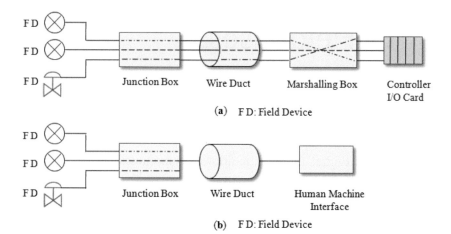

FIGURE 5.2 Schematic of (a) point-to-point and (b) fieldbus communication system.

TABLE 5.1
Comparison between a 4-20 mA System and a Fieldbus System

Serial number	Variable	4-20 mA system	Fieldbus system
1	No. of field devices per wire	1	Max. 32
2	Signal/Data	1	Up to thousands
3	Power supply over the wires	Yes	Yes
4	Signal degradation	Yes	Managed with terminators
5	Failure analysis	By human intervention	Reported at HMI
6	Maximum run length	2000 m with proper cables and power supply	1900 m, extendable to 5700 m with repeaters
7	Field device interchangeability	Yes	Yes
8	Intrinsic safety	Yes, more barriers needed	Yes, less barriers needed
9	Control in the field	No	Yes
10	Device failure notification	Very limited	Extensive
11	Networking of field devices	No	Yes

5.1.1 Evolution

With digital communication making its way into process automation systems several decades back, different vendors started developing their own protocols—independent of each other. In the initial stages of fieldbus introduction, design engineers were confronted with several problems. First, a particular vendor could not provide all the parts/components needed for a plant and that a particular manufacturer cannot make all the devices better than others. This led to either choosing the less-than-the-best devices from a single manufacturer or else settling for choosing best devices from different manufacturers. The latter option would give rise to integratibility problem and lead to isolated islands of automation and consequent interoperability difficulty with devices from different manufacturers.

Initially when fieldbus was introduced, it suffered from numerous problems like proprietary protocols, slow transmission speed and different data formats. Improvements in field signal transmission technology resulted in increasing levels of decentralization.

In 1985, industry experts in the field sat together to work out a vendor-independent fieldbus standard—i.e., it would be interoperable. The bus standard would provide bus power, intrinsic safety, and the ability to communicate long distances over existing wires—the basic requirements for a process plant automation system.

Partly due to the myriad complexities of instrumentation automation systems and mostly due to the reluctance on the part of the manufacturers, a single standard protocol architecture is yet to be established. FOUNDATION Fieldbus and PROFIBUS are now the two most dominant fieldbus technologies which are ruling the process automation field. Devices embracing these two technologies cannot communicate with each other because of protocol mismatch and thus seamless interoperability is yet to be achieved.

5.1.2 Architectural Progress

Over the years, control system architecture and field signaling developed in tandem. As field signal transmission technology continued to evolve, increasing levels of system decentralization resulted. Initially it was pneumatic transmission, followed by individual 4-20 mA analog current transmission lines to the control room. The control signals, emanating from the control room, were again taken to the actuators in the field for controlling the process variables.

In direct digital control (DDC), the central computer housed in the control room executed all the controls for each and every process variable situated far off in the field. In case of failure of the central computer, there would be total failure in the control strategy and the whole plant operations would be at grave risk. This led to decentralization of control

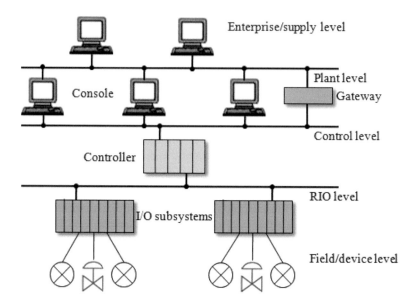

FIGURE 5.3 Multiple network architecture levels for DCS and PLC. (Courtesy: J. Berge. Fieldbuses for Process Control: Engineering, Operation and Maintenance, ISA, p. 20, USA, 2004.)

strategy and gave rise to a more decentralized distributed control system (DCS) and programmable logic controllers (PLCs) in early 1970s. Unlike the case of a single computer controlling the whole plant in the case of a DDC, in DCS several computers share between themselves the same task. In DCS, some 20–30 number of loops are controlled by a single computer. A single computer failure thus affects the loops assigned to that computer only. Traditional DCS and PLC architectures have multiple network levels as shown in Figure 5.3.

DCSs and PLCs emerged strongly with the advent of digital communication, but their architectures were based on 4-20 mA current output for field transmitters and valve positioners. In the field control system (FCS), there are only two networking layers—field-level network and host-level network. Thus, FCS architecture has taken the control much more into the field and it is much more decentralized than a DCS-based control system. FCS architecture is shown in Figure 5.4.

5.1.3 TYPES

There are many types of fieldbuses in use today; the particular type to be used depends on the type of industry—discrete or manufacturing automation. A list of the different types of fieldbuses is given below: FOUNDATION Fieldbus, PROFIBUS, DeviceNet, ControlNet, Interbus, HART, AS-i, Mod bus, CAN bus, Ethernet, Lon Works, WorldFIP, etc.

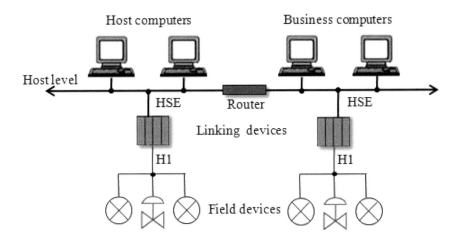

FIGURE 5.4 FCS architecture with control in the field devices. (Courtesy: J. Berge. Fieldbuses for Process Control: Engineering, Operation and Maintenance, ISA, p. 20, USA, 2004.)

5.1.4 Expanded Network View

Application of fieldbus in process automation has numerous advantages. Digital communication in case of fieldbus results in better control of a process along with higher product yield. Figure 5.5 shows a traditional process control with 4-20 mA current transmission scheme and a fieldbus-based

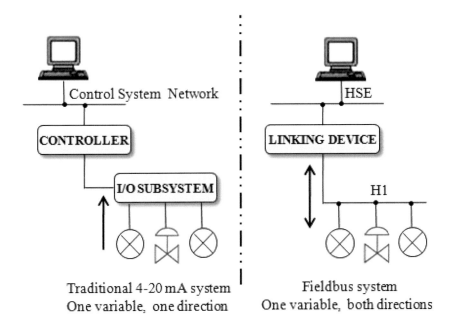

FIGURE 5.5 Traditional versus fieldbus.

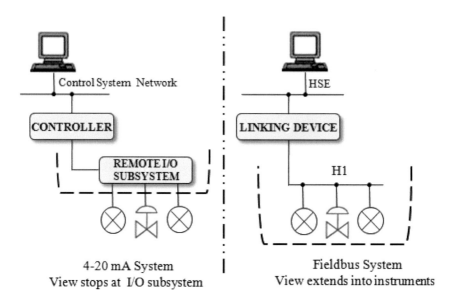

FIGURE 5.6   Extending the reach of fieldbus.

process control. For traditional control, a separate current transmission line is required for every process variable and it is one directional, while for fieldbus, many variables can be taken care of by a single transmission and it is bidirectional.

Figure 5.6 shows a comparison between a traditional and a fieldbus system in which the view is stopped at the remote I/O for the traditional one while it extends up to the sensors for the fieldbus system.

The I/O and control are all housed in the smart sensor in the field in case of fieldbus-based system, while a separate subsystem is required for the traditional system for which the controller is in the control room. Such a situation is shown in Figure 5.7.

Figure 5.8 shows a comparison between a conventional and a fieldbus-based system with regard to their applications in fire hazardous situations. For the former, for each device, an intrinsic safety (IS) barrier is needed while for the latter, a single IS is needed for many field devices.

5.2   TOPOLOGIES

Topology involves the manner in which the fieldbus devices are connected to the data highway. Several possible topologies are employed as per the needs of the plant geography. Several options employed are: point-to-point, bus with spurs (multidrop), tree or chicken foot, daisy chain, and mixed. For clarity, power supply and terminators are not shown in the figures representing different fieldbus topologies.

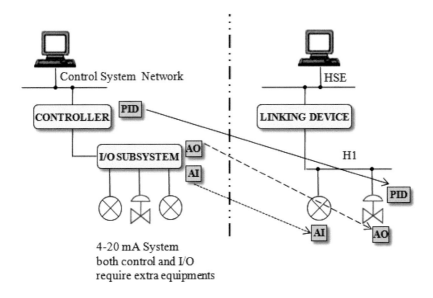

FIGURE 5.7 Control and I/Os in the field for a fieldbus system.

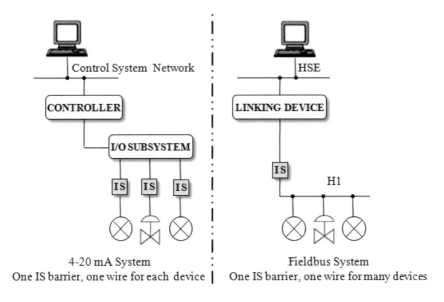

FIGURE 5.8 One IS barrier for many field devices.

5.2.1 POINT TO POINT

It is illustrated in Figure 5.9. In this topology, the segment consists of two devices. The two devices could be in the field or else one device in the field and the host in the control room.

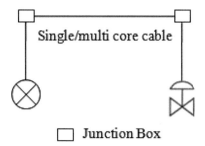

FIGURE 5.9   A point-to-point topology.

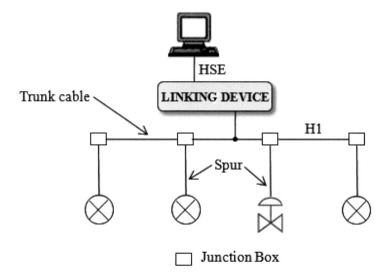

FIGURE 5.10   Bus with spurs or multidrop topology.

5.2.2   Bus With Spurs

A bus with spurs is shown in Figure 5.10, also known as multidrop. The devices are connected to the segment via small cable lengths, called spurs. Spur length can vary between 1 and 120m. The main cable spanning the whole length is called trunk cable.

5.2.3   Tree (Chicken foot)

In this, devices on a particular segment are connected via a junction box, marshaling panel, terminal, or I/O card (also called chicken foot). This scheme is suitable for devices situated in the same geographical place and can be connected to the same junction box and also obeying the rules of maximum spur length per segment. The scheme is shown in Figure 5.11.

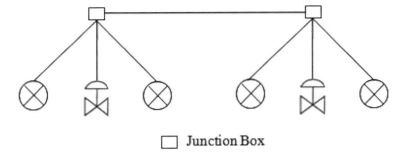

Junction Box

FIGURE 5.11   Tree or chicken foot topology.

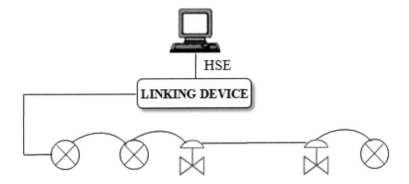

HSE

LINKING DEVICE

FIGURE 5.12   Daisy chain topology.

5.2.4   DAISY CHAIN

This is shown in Figure 5.12 in which the devices are series connected in a particular segment. Device connections to the segment should be such that disconnection of a single segment would not lead to total isolation of that segment.

5.2.5   MIXED

A mixed topology embraces more than one topology discussed above and a possible combination is shown in Figure 5.13. Depending on the physical locations of the devices in the plant and the length restrictions for segment lengths, different topologies are employed to derive the advantages of individual topologies.

5.3   TERMINATORS

On either side of a segment, a terminator (T) is placed, which acts as an impedance module and has the same characteristic impedance of the line.

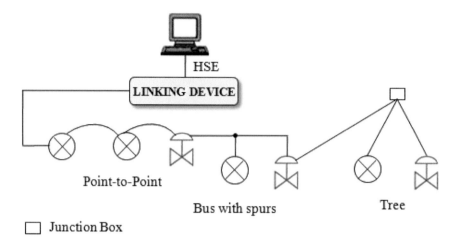

FIGURE 5.13   A mixed topology.

Terminators do not communicate, but helps in communication so that receiver error can be reduced to a minimum. A terminator prevents signal reflection and consequent communication errors. A terminator is a series combination of 100 Ω resistor and a 1 μF capacitor placed across the line. A fieldbus device transmits by changing its own current above a base level. The terminator converts this current change into an equivalent voltage change which is then used by the processing circuitry.

A single terminator is sometimes used for robust fieldbuses, but again it is susceptible to sporadic noise. Deployment of more than two terminators per segment again may lead to reduction in signal strength and consequent failures. A network, which cannot be accommodated in a single segment, is realized by bridging the different segments by repeaters. Thus, on either side of a repeater, two terminators are placed—one each for adjacent sides of two segments.

5.4   FIELDBUS BENEFITS

Fieldbus technology is applied in planning, installation, operation, maintenance, and renovation phases. In the planning phase, total plant integration by digital communication employing fieldbus can be achieved. In the installation phase, huge reduction in cost and time is achieved by replacing the traditional one-to-one wiring scheme with networking or multidrop communication. Optimum control due to enhanced functionalities in the field devices is derived in the operational phase by asserting control in devices. Self-diagnostic/calibration can be done *in-situ*—by software functionality in the maintenance phase. In the renovation phase, since network-based systems are modular—hence upgradation cost and time would be minimum.

Control action in fieldbus is distributed in the field devices. The field devices are smart and can carry out their own control/maintenance/diagnostics/calibration. The fieldbus system is intelligent enough to report its own failure and manual calibration can be done, if needed. By employing fieldbus-based control, efficiency in plant operations is enhanced and downtime reduced. Fieldbus is a not a product but a technology and belongs to the lowest level in the hierarchy, i.e., in field devices and exchanges information with the higher levels.

Since fieldbus devices are interoperable, interoperability does not become an issue and thus devices from different manufacturers can work together without any loss of functionality which was not the case with proprietary protocols. This has become possible because of standardization of specifications of fieldbus instruments leading to interoperable field-level networks. Fieldbus has eliminated the need for I/O subsystems and has the capability to detect/identify/assign addresses to devices.

Installation cost wise, there is huge savings of around 80 percent in wiring over conventional systems. Device commissioning time is greatly reduced because of considerable configuration and diagnostic information available in fieldbus devices. A field device can be diagnosed for its problems in *in situ* condition, thereby reducing downtime. Finally, since control has been shifted to the field devices, better and less complex central control system is required.

6 Highway Addressable Remote Transducer (HART)

6.1 INTRODUCTION

HART, an acronym for highway addressable remote transducer, is an open process control network protocol and was introduced in the late 1980s. It is a hybrid communication protocol which uses Bell 202 frequency shift keying (FSK) technique to superimpose digital communication signal on top of the analog 4-20 mA current loop signal. It is supported by HART Communication Foundation (HCF). Unlike the other *open* digital communication technologies applied to process instrumentation, HART is compatible with existing systems.

For many years, process automation and control industry is dominated by 4-20 mA analog signal to carry process variable and control signals to and from the control room. HART protocol extends this analog communication with additional bidirectional digital communication, being carried by the same wiring at the same time. This digital signal, known as HART signal, carries device configuration, diagnostic information, calibration, and any additional process measurements. Some of the features of HART protocol are as follows:

1. simultaneous analog and digital communication
2. compatible with conventional analog instrumentation schemes
3. supports multivariable field devices
4. flexible data access via up to two masters
5. open de facto standard
6. backward compatible
7. only protocol that supports both analog and digital, unlike other fieldbuses which are digital in nature
8. either point-to-point or multidrop operation
9. adequate response time of approx. 0.5 sec

6.2   EVOLUTION AND ADAPTATION OF HART PROTOCOL

HART was introduced by Rosemount in 1980s. Initially, the idea was to support the 4-20 mA loop current with digital process variable signals. Diagnostic, calibration, troubleshooting capabilities etc. of HART led to its more and more acceptance. This was even more so since the existing cables were utilized to carry the HART signals. Revision 4 was the first version that was introduced in the market having HART compatibility. Later, revision 5 was introduced in 1989 by Rosemount which has a huge installed base. It supports the latter version of the HART protocol.

HART is the only protocol that supports both analog and digital (hybrid) communication at the same time over the same pair of wires. Because of its widespread acceptability, the HART user group was formed in 1990 and HCF was formed in 1993.

6.3   HART AND SMART DEVICES

Developments in the fields of smart field devices and HART technology went on at around the same time. As more and more functionalities started emerging from the smart or intelligent field devices, more demands were placed from HART technology to compliment them. This led to further developments in either of them. Smart instruments are connected to programmable logic controllers (PLCs), computers etc. which collect sensor, actuator, or field instrument data and diagnostics by digital communication techniques.

Most smart instruments have the following capabilities: zero, span, and range adjustments, diagnostics to verify health, memory to store status and configuration information, etc. They support two way communications with the controller; digitize the process signal and process variables are corrected digitally. They contain advanced software which perform measurements and control actions are taken accordingly. Smart instruments add more resolution, more accuracy and above all more reliability to the process. At the site, various diagnostics can be verified in a few minutes by a handheld controller associated with HART technology.

Initially, HART was developed with primary process variable (PV) in mind. Later on, PVs included upper and lower transducer limits, etc. Status information was later included to lend more functionalities.

Smart field devices of todays are almost always multivariable in nature which is supported by HART technology. For example, a flowmeter includes a totalizer and a data logger. Initially, system manufacturers were slow to exploit the full capabilities associated with HART, but with time came the realizations and today, the huge installed base of HART compatible multivariable field devices are fully utilized to support them.

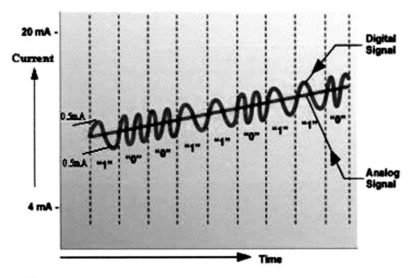

Digital over Analog

Note: Drawing not to scale

FIGURE 6.1 HART digital signal superimposed on 4-20 mA analog signal. (Courtesy: fieldcommgroup.com/technologies/hart/hart-technology-detail)

6.4 HART ENCODING AND WAVEFORM

HART protocol makes use of FSK technique to superpose the digital communication on top of 4-20 mA signals. The digital FSK signal is phase continuous and has a 0.5 mA amplitude having two different frequencies, 1200 and 2200 Hz, representing binary 1 and 0, respectively. Since the average DC value of a sine wave over a time period is zero, the FSK signal does not add any DC component on the 4-20 mA analog signal. Figure 6.1 shows the digital HART signal superimposed on the 4-20 mA analog signal.

The HART field devices and the central controller have FSK modems—the FSK modulated HART signals are carried along with 4-20 mA analog signal via the same wires. At the controller, HART signal is demodulated and actions taken accordingly.

6.5 HART CHARACTER

HART uses an asynchronous mode for communication purpose. Thus, HART data are transmitted one byte at a time without any clock signal. A HART character is composed of eleven bits and shown in Figure 6.2.

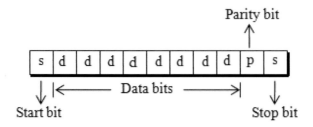

FIGURE 6.2   A HART character.

6.6 ADDRESSING

HART addressing is of two types: polling address and unique identifier. Polling address is single byte and is also known as *short address*. Unique identifier is of five bytes and also called *long address*.

The address field formats for the short and the long frames are shown in Figure 6.3. One bit of the short address distinguishes the two masters while another bit indicates burst mode telegrams. The rest four bits distinguish the field devices (from 0 to 15)—0 for single-unit mode and 1–15 corresponds to multidrop mode. The polling address format is used with old HART devices that do not support the long address format.

The five byte unique identifier is a hardware address which consists of one byte manufacturer code, one byte device type code, and a three byte sequential number. This five byte ID is unique for each device. HCF administers the manufacturer code which eliminates the possibility of address duplication of any two HART devices. The master uses this unique long address to communicate with the slaves.

In single mode, the master polls address 0 to get the unique slave ID. In multidrop mode, the master checks all the polling addresses 0–15 to check device presence. The master then presents a list of live devices on the network. A user can alternatively enter the tag of the intended device

Master	Burst	0	0	bit 3	bits 2 to 0

(a) Short frame format

Master	Burst	0	0	bit 3	bits 35 to 0

(b) Long frame format

FIGURE 6.3   HART address field formats: (a) short frame (b) long frame.

and the master will broadcast the same. The slave with the unique ID and the tag responds against this query from the master. The polling address, in conjunction with the unique ID, indicates whether the message exchange is from a primary or secondary master and whether the slave is in the burst mode or not.

6.7   ARBITRATION

Arbitration ensures proper message transmission between master and the slave devices. There can be either master/slave or burst mode. In the former, master initiates message transmissions by requesting the slave device. The slave, in turn, responds only to the query from the master. A slave in the master/slave mode of operation can never initiate communication.

There can be two masters on the network—a primary master and a secondary master—which may typically be a handheld terminal. Arbitration between the two masters is based on timing.

Slave burst mode, like the master/slave mode, is initiated by a command from the master. In this, burst mode responses are generated by the requested slave without request frames from the master. Data is updated at a faster rate since the slave goes on transmitting without a request from the master.

A frame, either from a master or a slave, is transmitted only after ensuring that no transmission is taking place on the network at that point of time. It is the responsibility of the timer which allows access to the network to the primary masters, secondary masters, slaves or slaves in burst mode. Both the masters have equal priority in getting access to the bus. In case both masters have to repetitively access the bus, they would do so alternately. Burst mode slaves wait longer than the masters to transmit, allowing the masters to control such slaves—either to continue or abort the burst mode.

6.8   COMMUNICATION MODES

Communication employing HART protocol can either be master/slave or burst mode. A HART communication loop may have two masters: primary and secondary. The master is typically a system host—may be a distributed control system (DCS), a PLC, or a personal computer (PC). The secondary master can be a handheld configuration tool (i.e., a handheld terminal) used for occasional configurations of different process parameters. A slave may be a transmitter or a valve positioner. Burst mode configuration provides the master with certain information on a continuous basis, until told to stop. It provides for a faster communication than the master/slave mode and is used in single slave configuration.

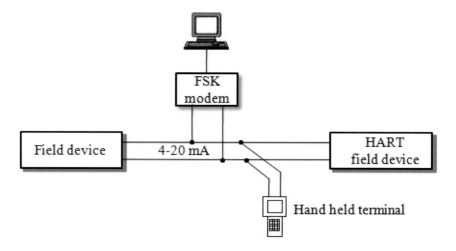

FIGURE 6.4   Point-to-point network configuration.

6.9   HART NETWORKS

HART networks can operate in two configurations: point-to-point and multidrop mode. In the point-to-point network shown in Figure 6.4, the traditional 4-20 mA current signal is used to control the process and remains unaffected by HART signal. The configuration parameters are transferred digitally over the HART protocol.

HART multidrop communication networks are used when the devices are widely spaced. As shown in Figure 6.5, only two wires are required to communicate with the master. If required, IS barriers and auxiliary power

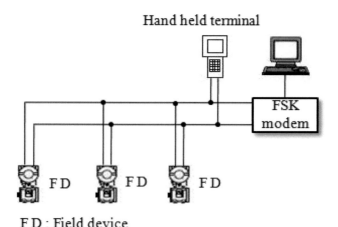

FIGURE 6.5   Multimode network configuration.

supply for up to 15 devices can be incorporated in this mode. All polling addresses of the devices are set at greater than zero and device current limited to a typical value of 4 mA.

Multidrop networks allow two wire devices to be connected in parallel. Information reading time from a single variable is typically 500 ms and with 15 devices connected to the network, approximately 7.5 seconds would be required to completely go through the network cycle once.

6.10   FIELD DEVICE CALIBRATION

HART devices need occasional calibration to ensure that the transducer output, via the different processing blocks, is representing the true process value. Any HART field device has transducer block, range block, and data acquisition (DAQ) block for such calibration.

Calibration of such field devices include the calibration of the digital process value, its proper scaling, converting into some desired range, and finally sending the 4-20 mA current value. The process output successively goes to the transducer block, range or range conversion block and finally to the DAQ block.

Figure 6.6 shows the calibration steps for an HART enabled field device. The transducer block involves comparing a simulated transducer value with an internally generated traceable reference. This comparison determines whether calibration of the field device is required, which can be done by using the HART protocol. Calibration is usually performed by providing the field device with exact transducer values—one near the lower limit and the other near the upper limit. Using HART protocol, the field device then performs the necessary adjustments, if needed.

The range block uses the lower and upper range values of 4 and 20 mA, respectively, to convert these transducer values into lower and upper range percent values—the former refers to 0 percent and the latter 100 percent.

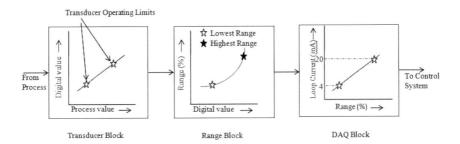

FIGURE 6.6 Calibration of HART field devices. (Courtesy: B. G. Liptak. Instrument Engineers' Handbook, Process Software Digital Network, 3rd Edition, CRC Press, Boca Raton, FL, p. 554, 2002)

The range block output may be in linear or square root form—as in the case of flow type process variable.

The output of the range block is passed on to the DAQ block which then converts the percent range values into loop current signals. Calibration of current loop is not very seriously needed because no moving parts are involved. Calibration at 4 mA corresponds to current zero trim and 20 mA called the current span trim.

6.11   HART COMMUNICATION LAYERS

The HART protocol follows the seven layer open system interconnection (OSI) protocol, although it uses only three layers: application, data link, and physical. The HART and the OSI protocol layers are shown in Table 6.1.

6.11.1   PHYSICAL LAYER

HART uses FSK physical layer which is based on Bell 202 modem standard. This kind of modulation is robust and has very good noise immunity. A HART modem chip is used at both the sending and receiving end sides for modulation and demodulation, respectively.

HART devices support both the conventional 4-20 mA current signal and modulated HART communications. These occupy different communication bands and shown in Figure 6.7. Because of their non-interfering nature, both communications are possible simultaneously. HART communication signal is filtered out by the analog devices and as such they remain unaffected by the HART signal. Thus, devices with 4-20 mA input or output work nicely in control loops.

TABLE 6.1
HART Protocol Layers Compared With OSI Layers

OSI layers	HART layers
Application	HART commands
Presentation	
Session	
Transport	
Network	
Data link	HART protocol rules
Physical	Bell 202

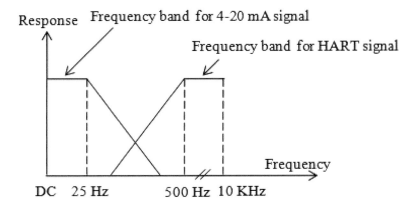

FIGURE 6.7   Frequency bands for 4-20 mA and HART signal. (Courtesy: B. G. Liptak. Instrument Engineers' Handbook, Process Software Digital Network, 3rd Edition, CRC Press, Boca Raton, FL, p. 554, 2002)

6.11.2   DATA LINK LAYER

The HART message frame format, often called a HART telegram, is shown in Figure 6.8. It consists of nine fields.

Preamble is the first field in the message which is sent first and it wakes up and synchronizes all the receivers of the connected devices in the network. The delimiter, a single byte field, signifies the end of the preamble. The delimiter is a start field and its content denotes if the frame is a request from a master, a response from a slave or a request from a slave in the burst mode. It also indicates whether the address used is a polling address or a unique ID.

The address field that follows may be a single byte (short frame format) for polling address or five bytes (long frame format) for unique ID address and discussed in Section 6.6.

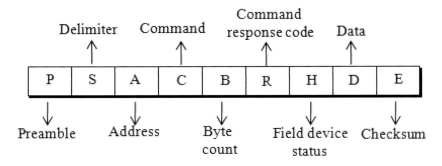

FIGURE 6.8   A HART character.

The fourth field is the command field and is of single byte. It represents the HART command associated with the message. Data link layer does not interpret the same but passes (accepts) the same to (from) the application layer. The byte count is of single byte length and indicates how many more bytes are still left for the message to be completed, excluding the checksum. The receiver can thus check for the end of the message from the information provided by this field.

The response code is also single byte and is included only in response messages. This response field from the slave would indicate the type of error occurred in the received message, if it was received erroneously. Correct message reception would also be indicated by this field. The field device status, a single byte field, is included in response messages to indicate the health of the device.

The data field may contain from none to 24 bytes. Data is not interpreted in the data link layer and is merely passed to the application layer. The checksum is of single byte and is longitudinal cyclic redundancy check (CRC) type. If data is received correctly, the transmitting end checksum value would be identical to the receiving end checksum.

6.11.3 APPLICATION LAYER

HART commands are defined in the application layer of the HART protocol. The communication routines of the HART master devices and the programs are based on these commands. Commands from the master can seek for data, start-up service, or diagnostic information. The slave, in turn, responds by sending back the required information to the master.

The HART *command set* includes three types of commands: *universal commands*, *common practice commands*, and *device specific commands*. The host application may implement any of the command types for a particular application.

6.12 INSTALLATION AND GUIDELINES FOR HART NETWORKS

HART devices use the conventional 4-20 mA current signal cable path to send the HART FSK signal. Thus, no extra communication paths are required to send HART signals. There are several ways to calculate HART cable lengths which depends on the quality of cables used and consequently on the capacitance—the lower the cable capacitance, the longer would be the cable run. Capacitance limitation imposed by intrinsic safety considerations further puts a limit on the maximum cable run in such cases. It is seen that HART communication is not vulnerable even for much degraded cable quality. Cable length is dependent on its size (cross section) and also whether it is single or multicore cable.

6.13   DEVICE DESCRIPTIONS

The universal and command-practice commands ensure interchangeability and openness of the field devices irrespective of the manufacturer. This is so when the user needs only standard status and the fault messages. The universal and command-practice commands become insufficient if some device-specific information or some special properties of a device are required. The software of the master device has to adapt itself for such occasional device-related information or a new device is included in the system. This is where the device description language (DDL) was developed.

DDL is not limited to applications related to HART only, but can be applied to other fieldbuses also. DDL is like a programming language which enables device manufacturers to describe all communications in a complete and comprehensive manner. It is a powerful tool that provides significant benefits during configuration and device specific features.

DDL supports all extensions. Thus, any device or manufacturer specific command can be executed by DDL and the user is provided with a universally applicable and uniform user interface. This thus entails a clear, user friendly and safer operation and monitoring of the process.

A DDL encoder reads the device description as binary encoded DD data and which can be read by a master device. If a device has sufficient storage space, the said DD can be stored in its firmware. During the parameterization phase, it can be read by the master device.

6.14   APPLICATION IN CONTROL SYSTEMS

Because a combination of analog signal and digital data are available at the same time from HART enabled multivariable field devices, a multitude of data—both for the process variable, health of the process, and environmental conditions can be monitored at the same time. HART field devices can update their data twice or thrice in a second. Although it is not very fast, this rate of data updation is much faster than the time constraints associated with most of the processes. A HART multivariable pressure transmitter may have a temperature sensor which can monitor the process temperature. This sensor output would warn against any damaging temperature condition thereby eliminating the need to employ separate temperature sensor for the above. This saves time, money, and space.

In the legacy 4-20 mA current transmission scheme, the field wiring is brought to a junction box. A multicore *home run* cable connects the junction box to the I/Os in the marshalling room. This is then connected to the controller via a backbone bus. The controller provides data to operator consoles, engineering workstations, etc. In this traditional scheme, only information is in the form of an analog 4-20 mA signal.

Latter architectures employ remote I/Os, operator consoles, controllers, engineering workstations all connected by an *open* communication protocol. The remote I/Os are located close to the processes in which the existing cables are connected to HART field devices. The multicore cable is thereby dispensed with network cables providing two-way HART communication which enhances both faster data availability and integrity.

Another major area in which HART enabled systems score over the traditional ones, is the availability of status information. This status information can be sent via the network cables to the control room. Thus, an *apriori* knowledge about plant's possible breakdown is available and preventive maintenance can be undertaken—reducing outages and downtime.

6.15   APPLICATION IN SCADA

The technology used in HART is very useful for SCADA applications. Since HART supports digital process values, it can be applied in SCADA. Various status parameters of process variables obtained from HART measurements can be used to schedule field trips for preventive maintenance. This would result in less downtime and less employment of manpower.

SCADA applications require updating field data in minutes. Since data updating using HART requires only a fraction of a second, it can very safely be applied in SCADA applications. Other application areas where HART can be applied in SCADA are: inventory management, automated meter reading, pipeline monitoring situated in far off places, and remote monitoring of remotely located petroleum production units.

6.16   BENEFITS

HART protocol is a unique communication facility that is backward compatible ensuring that existing investments in plant cabling and power needs remain secure in future also. It is the only communication facility that supports both analog and digital communication (hybrid) at the same time. By preserving the traditional 4-20 mA signal, it extends the system capability for two way digital communication with multivariable smart field instruments. Any process application can be addressed by the services offered by HART protocol.

The simplicity of HART protocol and its low-cost HART–compliant field devices is the basis for its widespread acceptability in process instrumentation systems. Other advantages that accrue by employing HART protocols in process industries are: fully open de facto standard, multiple smart devices along the same highway, multiple masters can control the same smart device, long distance communication via telephone lines—can be implemented by using a HART-CCITT converter, up to 256 process variables belonging to the same device can be taken care of, addition of new features without much difficulty, access to device parameters, and diagnostics.

7 Foundation Fieldbus

7.1 INTRODUCTION

The Foundation Fieldbus has its origin in Fieldbus Foundation—a non-profit consortium that recommends and develops the technical specifications for Foundation Fieldbus. These are based on ISA/ANSI S50.02 and IEC 61158. Fieldbus Foundation was formed in 1994 with the merging of ISP (Interoperable Systems Project) and WorldFip North America.

The traditional analog 4-20 mA current signal is based on ISA S50.1. The ISA standard committee SP 50 met in 1985 to develop a standard to replace analog current transmission with a digital communication link for instrumentation and automation sector. It was the need of the hour to develop a common standard—from the economic and technical aspects—so that best instruments from different manufacturers can be put together to run a plant without facing any interoperability problem.

The standards committee recommended and defined two different network requirements: a fairly low speed H1 was recommended to be installed at the plant floor level (sensor level) to replace the 4-20 mA current transmission retaining the existing plant wiring. Second, a higher speed H2 was recommended which acts as a backbone to the H1 segments. Conformance testing for H1 was completed in 1998 by the Foundation Steering Committee. The same committee, in 1999, recommended the implementation specifications for H2 based on commercial off-the shelf (COTS) high speed Ethernet operating at 100 Mbps. It was completed in 2000 and thus H2 was replaced by high speed ethernet (HSE).

7.2 DEFINITION AND FEATURES

Fieldbus Foundation defines "Fieldbus is a digital, two-way, multidrop communication link among intelligent and control devices" (Technical Overview, Foundation Fieldbus, FD-043, Revision 2.0 HYPERLINK "http://www.fieldbus.org" www.fieldbus.org). It is one of several local area networks (LANs) dedicated to industrial sector.

The features associated with Foundation Fieldbus are as follows:

1. It is a bidirectional, half-duplex, digital process control and automation protocol.

2. When compared with the open system interconnection (OSI) reference model, Foundation Fieldbus has three layers: physical layer (PHL), data link layer (DLL), and application layer (APL). It has an additional layer—layer 8, called the *user layer.*
3. It has a two-level architecture consisting of the lower-level H1 and the upper layer H2 (HSE).
4. It supports both scheduled and unscheduled communications.
5. It supports interoperability, i.e., devices from different manufacturers can be seamlessly connected.

7.3   FOUNDATION FIELDBUS DATA TYPES

Different data types are combined to form complex data objects. The different data types used in Foundation Fieldbus are: Boolean, integer, floating point, unsigned, octet string, visible string, date, time of day, time difference, time value, bit string, null and packed.

7.4   ARCHITECTURE

Foundation Fieldbus has a two-level architecture: H1 is the lower-level bus that connects the field devices together and HSE is the upper-level bus connecting different H1 bus segments. The former operates at a speed of 31.25 kbps while the latter at 100 Mbps. The two-level Foundation Fieldbus architecture is shown in Figure 7.1.

The architecture supports control in the field devices such as control valve positioner and field instruments. It supports total control in the field

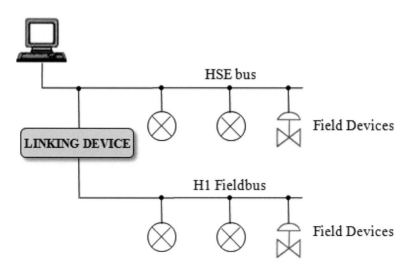

FIGURE 7.1   A two-level Foundation Fieldbus architecture.

(field devices) with no control room equipment. It also supports cascade control linking field devices and control loops in dedicated controllers.

Both H1 and H2 run on the same protocol and perform identical services. But their physical and data link layers are quite different. The two-level architecture of Foundation Fieldbus is made possible with the implementation of a *user layer*—sometimes called *layer 8*. The building block of the user layer lies in the *function block*.

7.5   STANDARDS

ISA (Instrumentation, Systems, and Automation Society) recommends standards which operate under ANSI rules. The Foundation Fieldbus has its standards based on ISA/ANSI S50.02 and IEC 61158. These two standards recommend communication specifications, requirements, and methods for Foundation Fieldbus. S50.02 came into being in 1993, while the standard IEC 61158 was introduced in 1999. The physical, data link, and application layers belonging to ISO 7498 of the seven layer OSI are fully specified in this standard.

The Foundation Fieldbus specifications are also compliant with IEC 61804 (for Function Blocks for process control and Electronic Device Description Language) and IEC 61508 (for electrical/electronic/programmable electronic safety related systems).

7.6   H1 BENEFITS

Application of H1 technology offers significant benefits in the control system life cycle which are: a huge reduction in the number of wires and marshalling panels, reduced number of I/Os, reduced number of IS barriers, reduced control room size, reduced number of equipment, reduced number of power supplies, reduced downtime, etc. Other benefits include ability to configure the devices remotely, increased accuracy in measurements, increased information availability with regard to health and status of equipment, *in-situ* calibration, increased plant safety, etc.

Multivariables from each device can be brought to the control room for trend analysis, archival purposes, predictive maintenance, asset management and report generation, etc. An expanded view of the system right from the control room level to the field level is possible by employing H1.

7.7   HSE BENEFITS

Different plant systems are integrated by HSE through control backbone. Resource allocation for maintenance can be assigned via asset management with the help of HSE. The corresponding layers of H1 and HSE are

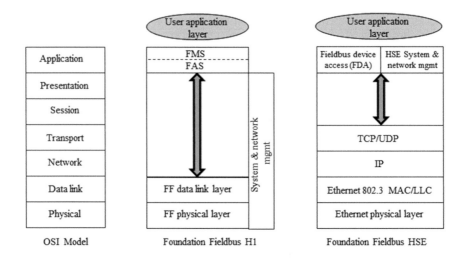

FIGURE 7.2 The two Foundation Fieldbus levels H1 and HSE compared with OSI layers.

compared with OSI layers and shown in Figure 7.2. The benefits of having HSE in the upper tier of the architecture are discussed in the following subsections.

7.7.1 Interoperability of Subsystems

A plant has different subsystems. A blast furnace, for instance, has weighing system, excess gas bleeder station, charging of different inputs to the blast furnace, tapping of liquid hot metal, and slag intermittently. A furnace operator can integrate these operations because of open protocol. Using HSE, information for different plant operations can be accessed without customer programming. HSE's help is also taken for data integrity, device health checking, redundancy, etc.

7.7.2 Function Blocks

Function blocks are identical for both H1 and HSE devices. Foundation fieldbus eliminates the need to have proprietary programming language. Instead, the same programming language is used for all plant operations. Function blocks are responsible for any loop shutdown, bumpless transfer, and the like.

7.7.3 Control Backbone

Linking devices bring data from individual segments (belonging to H1) to the HSE control backbone. Thus, different H1 networks are bridged by

HSE. Control backbone has total control starting from individual processes up to the different plant areas spread across the total plant territory. HSE dispenses with the use of remote I/O networking level, enterprise and control level, thereby flattening the enterprise pyramid.

HSE supports peer-to-peer communication capability. A device can communicate with another one without having to go through the central computer—thus eliminating any risk in case of central computer failure.

7.7.4 Standard Ethernet

HSE uses standard Ethernet cables, interface cards, and networking hardware which are available very cheap. Ethernet communications can be wireless, fiber optics, or twisted pair. Networking hardware from different suppliers are available both in commercial and industrial grades. Ethernet components are available in standard COTS category.

7.8 THE COMMUNICATION PROCESS

The communication process involves transportation of data and messages from one fieldbus device to another. The devices may reside on the same link or other links joined by bridges. Devices must have a device tag for proper device identification and a device network address for properly delivering data at the designated device. Communication between a sensor and an actuator (both field-level devices) is shown in Figure 7.3.

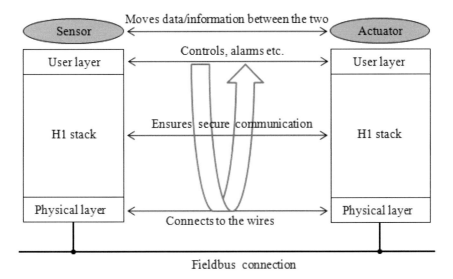

FIGURE 7.3 Communication between a sensor and an actuator.

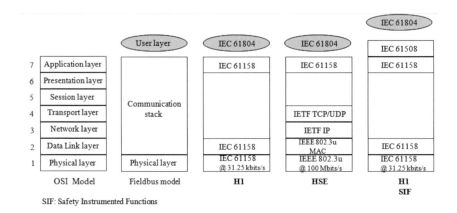

FIGURE 7.4 Fieldbus model compared with OSI reference model.

7.8.1 OSI Reference Model

Foundation Fieldbus H1 technology is based on OSI reference model. Figure 7.4 shows a comparison between OSI reference model and Fieldbus model.

The Fieldbus model consists of three layers: physical layer (PHL), data link layer (DLL), and application layer (APL). For Fieldbus model, application layer is divided into two sublayers: fieldbus message specification (FMS) and fieldbus access sublayer (FAS). FAS maps FMS into the data link layer. There is a layer above layer 7—called *user application layer*. This is the layer 8 of the Fieldbus model, although it is absent in OSI reference model.

In the Fieldbus model, layers 3–6 of OSI are not used. Layers 2–7 are mostly implemented in software and termed as the *communication stack*.

7.8.2 The PDU

Figure 7.5 shows how user data is passed down the layers to reach the lowest level, i.e., physical layer. From the physical layer it is then put on the fieldbus. It is seen that each layer appends protocol control information (PCI) to the message that it receives from the higher layer. The total message that a layer shifts to its just lower layer is known as the protocol data unit (PDU) of the former. As an example, FAS layer contains the FMS PDU (this is the total message received from the FMS layer) along with the PCI of FAS.

7.8.3 Physical Layer

As shown in Figure 7.5, physical layer receives the DL PDU from the data link layer to which are appended preamble, start delimiter, and end delimiter. This is then converted into physical signal and transmitted on the

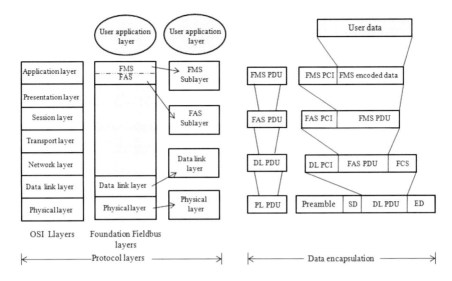

FIGURE 7.5 Concept of PDUs in fieldbus model.

Fieldbus medium by electrical or optical means. Physical layer interests are signal and waveform shapes, voltage levels, type of wires, etc.

On the physical wires, data at the rate of 31.25 kbps are sent. Although the speed of data transmission is not very high enough, it serves the purpose of automation industries in the instrumentation sectors with the cabling remaining intact. Explosive gases present in some areas in the plant prevent electronic devices to be placed there, and 31.25 kbps transmission fits into nicely which demand devices with low power consumption.

7.8.3.1 Manchester Coding

Synchronous serial communication in half-duplex mode is used in Foundation Fieldbus. Manchester Biphase-L coding scheme is used to code data generated from the field devices and transmitted along the bus. In this coding scheme, there is always a change in the coded pattern at the mid-point of each clocking, irrespective of data pattern.

7.8.3.2 Signalling

A typical physical layer signal waveform is shown in Figure 7.6. The pre-amble is sent first, followed by the rest in the sequence shown. Preamble is an 8-bit sequence of alternating 1's and 0's and is used to synchronize the receivers on the bus. Start and end delimiters are also 8 bit in lengths. In both start and end delimiters, N+ (non-data positive) and N- (non-data negative) symbols, which are high- and low-level signals span the whole clock duration. At the receiver, the start delimiter is used to identify the start of data bits and the end delimiter is used to signal end of data.

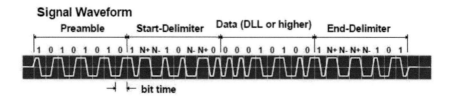

Signal Waveform

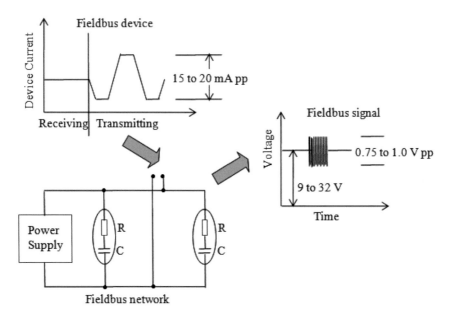

FIGURE 7.6 Physical layer signal waveform. (Courtesy: Yokogawa Electric Corporation Foundation Fieldbus Book-A Tutorial, Technical Information TI38K02A01-01E, 2000.)

FIGURE 7.7 Voltage waveform on the bus for a transmitting fieldbus device. (Courtesy: Technical Overview-Foundation Fieldbus. www.fieldbu.org)

The Fieldbus devices are supplied with a DC voltage in the range of 9–32 V. The device which is transmitting on the bus does so by delivering ±10 mA at 31.25 kbps into a 50 Ω equivalent resistor. This develops a 1.0 V peak-to-peak voltage modulated on top of the DC supply. Figure 7.7 shows the voltage wave shape on the bus when the device is transmitting.

7.8.4 DATA LINK LAYER

Data link layer manages transfer of data from one node to another. It also manages priority of data transfer amongst the field devices. Addresses, data transfer, their priority, medium control are all managed by data link control layer. It prevents access to the bus by two or more devices present in the segment at the same time.

Data link layer supports transmitting data as per urgency. There are three levels of urgency: URGENT, NORMAL and TIME_AVAILABLE, in this order. An urgent message overrides all other message queues. Maximum data size belonging to the three data types are 64, 128, and 256 bytes, respectively.

7.8.4.1   Medium Access Control (MAC)

The most important job of data link layer is medium access control (MAC) of the fieldbus devices. A link or segment is the particular part which shares the physical layer signal at the same time. A device, at a given instant of time, is allowed to use the link by proper software programming. A link active scheduler (LAS) controls this medium access in the data link layer.

7.8.4.2   Addresses

There are many devices in a link and many links are there which are connected by bridges. Devices on a fieldbus are identified by DL-address. It consists of three fields: link, node, and selector, having their respective lengths 16, 8, and 8 bits. A link address identifies a link and if communication is between the same link, it is omitted. The content of the node field gives the address of the node in a link. The selector field gives device—internal 8 bit address. It identifies a virtual communication relationship (VCR). When a VCR is connected to another VCR, it is identified with data link connection end point (DLCEP). Data link service access point (DLSAP) is used when a VCR is not connected to another VCR, but can send/receive messages.

7.8.4.3   Link Active Scheduler and Device Types

Foundation Fieldbus devices are classified into basic, link master (LM), and bridge types. Basic class devices cannot become LAS, while LM class devices can. A bridge class device can become a LAS in addition to its original functionality to connect different links.

In a link there can be more than one LM, but only one can act as LAS at any given time. When there are no LAS in a link, LM devices try to acquire the LAS role, but the one with the least node address wins this contention. Figure 7.8 shows how a LM class device becomes a LAS.

7.8.4.3.1   Scheduled Communication

Scheduled data transfers are used for regular cyclic transfer of control loop data between devices residing on the Fieldbus. LAS controls the periodic data transfer from a publisher (source of data) to subscribers (data sink) using network schedule. Thus, schedule transfers belong to publisher/subscriber type of reporting for data transfer. Scheduled communication takes place in a synchronous manner.

LAS issues a compel data (CD) to the device when its turn comes to upload data on the bus. The device then broadcasts or publishes its data in

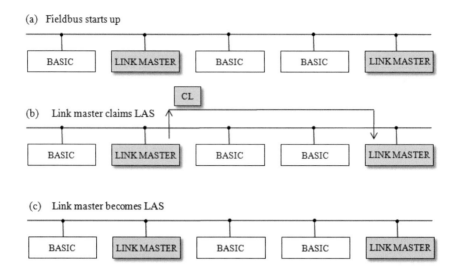

FIGURE 7.8 Process of a LM class device becoming a LAS. (Courtesy: Yokogawa Electric Corporation Foundation Fieldbus Book-A Tutorial, Technical Information TI38K02A01-01E, 2000.)

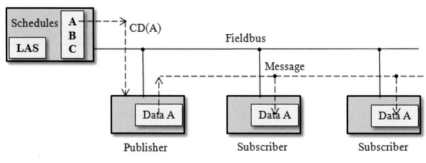

CD: Compel data LAS: Link active scheduler A, B & C: Data

FIGURE 7.9 Scheduled data transfer technique. (Courtesy: Technical Overview-Foundation Fieldbus. www.fieldbu.org)

the buffer of all the devices connected to the bus. The device configured to receive data is called a subscriber. The scheduled data transfer scheme is shown in Figure 7.9.

7.8.4.3.2 Unscheduled Communication

Unscheduled communication takes place in an asynchronous manner. LAS gives a chance to all the devices on the bus to send unscheduled messages between transmission of scheduled messages. LAS issues a pass token (PT) message to a device. This is shown in Figure 7.10. On receiving

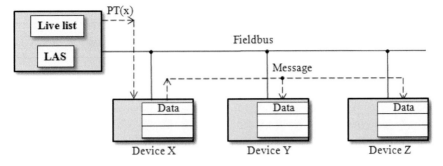

LAS: Link active scheduler   PT: Pass token

FIGURE 7.10 Unscheduled data transfer technique. (Courtesy: Technical Overview-Foundation Fieldbus. www.fieldbu.org)

the PT, the device transmits messages until it is finished or until the *delegated token hold time* has expired—whichever is less. The message can be sent to a single destination or multiple destinations. In the figure, message from device x is sent to both the devices y and z in a multicast manner.

Unscheduled data transfer uses either client/server or report distribution type of reporting for transferring data. Usually such transfer scheme is undertaken for user initiated changes like set point change, tuning change, mode change or upload/download.

7.8.4.3.3   LAS Operation

LAS operation can best be understood by the Link Active Scheduler algorithm shown in Figure 7.11. The flow chart shows the overall sequence of operations performed by LAS. If there is still some time left after

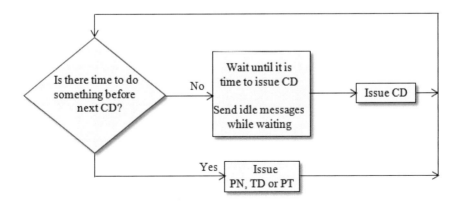

FIGURE 7.11 LAS algorithm. (Courtesy: Technical Overview-Foundation Fieldbus. www.fieldbu.org)

completing the allotted job on issuance of CD, the algorithm may issue a probe node (PN), time distribution (TN) or pass token (PT) as already programmed. Else, the system waits until it is time to issue the next CD. The system thus does the synchronous cyclic jobs interspersed with acyclic jobs assigned to it during programming.

7.8.4.3.3.1 CD Schedule   CD schedule contains a list of activities which are carried out by LAS in cyclic communication. The schedule is prefixed. Thus, at some precise point of time, LAS sends a CD message to a specific fieldbus device to put its data on the bus. The device then publishes its data on the bus and it goes to the subscribers. The CD schedule is the highest priority job of the LAS; the other jobs are carried out in between scheduled transfers.

7.8.4.3.3.2 Live List Maintenance   *Live list* refers to the devices on the bus which responds to the PT issued by LAS. Devices may be added/ deleted to/from the fieldbus as per requirement. A device may also go off the list if it goes bad. It is the responsibility of LAS to maintain and update the list of live devices on the list.

LAS probes at least one node (i.e., one address) after it has completed one cycle of sending PTs to all the devices residing in the live list. A device may be added to the bus at any time. LAS periodically sends a PN message to the addresses not on the live list. If a device is detected at the node, it immediately sends back a probe response (PR) message and the device is added to the live list by the LAS. LAS also confirms the device by sending it a node activation message.

LAS removes a device from its live list if it does not respond to the PT or returns the same to the LAS consecutively for three successive tries.

Whenever a device is added or removed from the live list, LAS immediately broadcasts the changes in the live list to all the devices on the bus. Thus, each link master on the segment maintains a current and updated copy of the live list.

7.8.4.3.3.3 Data Link Time Synchronization   A time distribution (TD) message is broadcast on the fieldbus by the LAS. This ensures that all the devices on the bus have exactly the same data link time. Scheduled cyclic communication on the bus and scheduled function block executions are based on such TD message information so that exact time synchronization is always maintained.

7.8.4.3.3.4 LAS Redundancy   A segment or link has more than one link master. In case one LAS fails, another link master will take over. This will ensure that communication on the bus is not disturbed. Thus, the fieldbus is designed to be *fail operational.*

7.8.5 APPLICATION LAYER

It consists of two sublayers: FAS and FMS. FAS manages data transfers while FMS encodes/decodes user data.

7.8.5.1 Fieldbus Access Sublayer

FAS provides services to FMS with the help of scheduled and unscheduled communications of the data link layer. Services provided by FAS are described by virtual communication relationships (VCRs).

7.8.5.1.1 VCR

A VCR can be thought of as equivalent to the speed dial feature on a memory telephone. For an international call to be established, international access code, country code, exchange code, and finally the specific telephone number are required. These details are entered in the memory of the telephone and a *speed dial number* is assigned to it. Whenever a call is to be established with this particular telephone number, only this speed dial number is entered for dialing to take place. Similarly, after configuration, only the VCR number is needed to be communicated with another fieldbus device.

Just as there are different types of calls, like calls to different individuals belonging to different countries, conference calls, call collect, etc., there are different types of VCRs to cater to various devices or applications at the same time. The VCR guarantees that message goes to the specific and correct partner.

When a message is transferred, it goes through the VCR by adding protocol communication interface (PCI) before going to the physical wire. A VCR is identified by a device-local identifier called *index* specified in the application layer. A VCR can also be identified from other devices with a *DL-address* specified in the data link layer. A VCR has memory to save messages.

Network management configures a particular VCR by providing it with correct index and DL-address. Thus, no two VCRs can have identical index and DL-address.

There are three different types of VCRs available: Client/Server VCR type, Source-Sink (Report Distribution) VCR type, and publisher/subscriber VCR type, shown in Figure 7.12.

7.8.5.1.1.1 Client-Server VCR Type

It is a user initiated one-to-one, queued, unscheduled, prioritized communication between devices on the fieldbus. Queued means that messages are received as they are sent, with message priorities remaining intact. In this method, messages are not overwhelmed as in publisher-subscriber type. Transfers in this VCR type is flow controlled and has a retransmission scheme in case of corrupted message transfers.

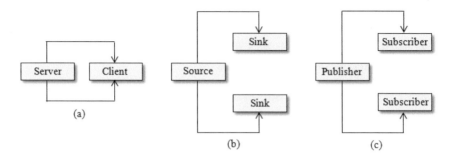

FIGURE 7.12 The three types of VCRs: (a) Client/Server VCR type; (b) Source-Sink VCR type; (c) Publisher/subscriber VCR type. (Courtesy: N. Mathivanan. PC-based Instrumentation: Concepts and Practice, PHI, p. 587, 2007.)

Operator initiated messages such as set point change, alarm acknowledge, tuning parameter change and access, device upload/download, etc. are done in client-server VCR type.

A device, when it receives a request from LAS via PT, sends a request message to another device. The requester is the client while device that received the request is called a server.

A typical example is a human machine interface (HMI) wanting to read data from a function block. The former is the client and the latter is the server.

7.8.5.1.1.2 Publisher—Subscriber VCR Type It is used for buffered, one-to-many communications to transfer process critical data such as process variable. Buffering means the last published data on the network is maintained and held until a new data overwrites it. Data producer is called a publisher and data receiver is called subscriber.

In this VCR type, data can be transferred on a purely periodic basis by properly scheduling CD in the LAS. Unscheduled transmission of information is also possible.

7.8.5.1.1.3 Source-Sink (Report Distribution) VCR Type It is a one-to-many, one-way communication without schedule. It is used to broadcast or multicast event and trend reports. The destination address may be predefined or included separately with each report. It is sometimes called report distribution VCR type.

Transfer in this mode takes place in a queued manner. Messages are delivered in the order in which they are transmitted. The transfers are unscheduled and take place in between the scheduled ones. This mode is typically used for by fieldbus devices to send alarm reports to operator consoles.

7.8.5.2   Fieldbus Message Specification

Applications use FMS services to send messages between themselves across the fieldbus using a standard set of message formats. FMS is a model for applications to interact over the fieldbus. Virtual field device (VFD) and object dictionary (OD) belong to this model. Message formats, communication services, protocol behavior are the services that FMS provides for the user application layer.

OD can be thought of as a look-up table that gives information about a value, such as data type, that can be read from or written into a device. Data communication over the fieldbus is described by an *object description*. Such descriptions are structured together in OD.

VFDs are used to remotely view local device data which are described in the object dictionary. An identifier, maintained in a VCR, identifies the VFD.

A Foundation Fieldbus device will have at least two VFDs—one is Management VFD and the other is function block VFD. A field device may have two or more function block VFDs. Network and system management functions reside in management VFD. Such a VFD configures network parameters that include VCRs. It also manages fieldbus devices.

FMS provides different services to access different FMS objects. These are: variable access, event management, context management, domain management, upload/download, program invocation, etc.

7.9   TECHNOLOGY OF FOUNDATION FIELDBUS

Communication in Foundation Fieldbus is based on OSI. It uses only three layers viz. physical, data link and application layer of the OSI. For proper operation of the fieldbus, it defines another layer which is called layer 8 and termed as the user application layer. This layer is standardized by Fieldbus Foundation based on blocks. The different blocks in the user layer are: resource block, function block, and transducer block. Devices on the fieldbus are configured by resource and transducer blocks while control strategy is built using function block.

Data transmission and reception on the fieldbus take place either by electrical wires or optical means at 31.25 kbps via the physical layer. This speed is sufficient to cater to the needs of instrumentation and control and their corresponding automation aspects.

The voltage supplied to the system can vary between 9 and 32 V. When a device transmits, it delivers ±10 mA at a data rate of 31.25 kbps into a 50 Ω resistor generating a 1 V peak-to-peak voltage. The DC supply voltage is thus modulated by this 1 V signal.

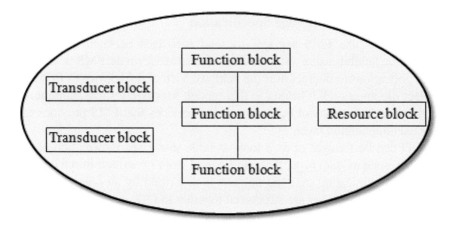

FIGURE 7.13   Blocks in the user layer.

7.9.1   USER APPLICATION BLOCKS

Layer 8, the user layer, is shown in Figure 7.13. The user layer defines blocks and objects which represent the functions and data available in a device. For Foundation Fieldbus, HMI is done through these blocks and objects—rather than through a set of commands as is followed in most communication protocols.

The blocks can be thought of as processing units. They accept inputs; their settings can be changed as per system requirements. They also have an algorithm, which when run, can produce outputs. The blocks also communicate with each other.

7.9.2   RESOURCE BLOCK

A resource block indicates the resources available with a device. This may include device type and revision, serial number, manufacturer ID, and resource state. A device can have only a resource block. Resource block controls the overall hardware of the device and function blocks within the VFD.

The resource block contains its hardware specific characteristics. It does not have any input or output parameter. The algorithm within the resource block monitors and checks the device hardware performance. It must always be in automatic mode. With its help, cause of any problem related to the device can be fixed and appropriate actions taken.

7.9.3   FUNCTION BLOCK

Function block(s) in a fieldbus device perform different functions required in a process control operation. A fieldbus device may contain one or more function blocks. In process control systems, operational requirements normally vary—depending on the particular process and the end product quality. Thus, Fieldbus Foundation has designed a standard function block as normally applicable in process industries. In addition, some other function blocks have been recommended. Most importantly, with the help of models and parameters existing in the function block, the design engineer can configure, maintain, and customize the applications.

A function block has input, output, and contained parameters. Data generated in a block is available at its output and acts as the input to another block. Thus, an AI block's output acts as the input to the proportional-integral-derivative (PID) block. Its output is the input to the AO block, whose output may be fed to the PID block for process control purpose.

Figure 7.14 shows a control loop which uses function blocks in fieldbus devices.

7.9.3.1   Function Block Library

The function block library is divided into several parts as defined by Fieldbus Foundation. In Part-2, there are ten basic function blocks which are the fundamental ones supported and required in process control and measurement. Thus, the parameters of such blocks must be known to all systems. Part-3 blocks are used for advanced measurements, while Part-4 blocks provide I/O interface to the outside world.

Attributes of a function block like data type, its name, minimum and maximum values wherever applicable are contained in DD—the device description block. For every function block, there is a DD—be it basic, extended, or custom type function block. The basic and extended function blocks are documented by Fieldbus Foundation in their respective function block specification.

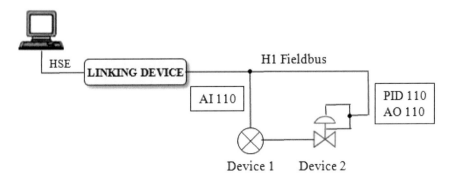

FIGURE 7.14   A control loop using function blocks. (Courtesy: Technical Overview-Foundation Fieldbus. www.fieldbu.org)

7.9.3.2 Function Block Scheduling

Figure 7.15 shows the scheduling of function blocks. A schedule building tool is used to generate the proper time schedules for operations of both LAS and different function blocks. The schedules have been built on the control loop operation shown in Figure 7.14. The schedule building tool contains start time offsets of the different function blocks from the *absolute link schedule start time.*

A *macrocycle* is defined as a single iteration of a schedule within a device. There can be device macrocycle, LAS macrocycle. Table 7.1 shows the time offsets of AI, AO, and PID blocks from the absolute link schedule start time.

AI block is executed at offset 0. At offset 20, LAS issues a CD to the AI function block buffer residing in the transmitter and then data in the buffer is published on the fieldbus. At offset 30, PID function block is executed, followed by execution AO function block at offset 50.

The different offsets represent the exact time at which a particular function starts its operation with respect to absolute link schedule start time. The beginning of the macrocycle represents a common and unified start time for all concerned function blocks connected to the link and for the

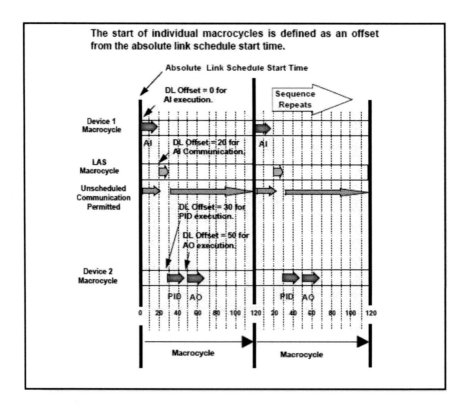

FIGURE 7.15 Scheduling of function blocks. (Courtesy: Technical Overview-Foundation Fieldbus. www.fieldbu.org)

TABLE 7.1

Schedules of Different Function Blocks

Schedule of different blocks	Offset values
AI execution	0
AI communication	20
PID execution	30
AO communication	50

Source: Yokogawa Electric Corporation Foundation Fieldbus Book-A Tutorial, Technical Information TI38K02A01-01E, 2000.

LAS link wide schedule. Different function block execution and their corresponding data transfers on the bus are thus synchronized in time.

It is apparent from Figure 7.15 that from offset 20 to offset 30, the bus cannot be used for sending unscheduled messages. This is so because during this time AI publishes (communicates) data on the bus. AI obtained this data during its execution from offset 0 to offset 20.

7.9.3.3 Application Clock Distribution

Foundation Fieldbus supports an application clock distribution function. It is usually set and adjusted to local time of the day or to some universal coordinated time. System management contains a time publisher that periodically sends an application clock synchronization schedule message to all connected fieldbus devices. The data link scheduling time is sampled and sent along with the application clock message. This allows the receiving devices to adjust their local application time. In between the synchronization time message, the application clock time is independently maintained in each device based on their internal clocks.

Application clock synchronization allows fieldbus devices to time stamp data throughout the fieldbus network. If a backup application clock publisher is maintained in the network, the same would become active when the currently active time publisher fails.

7.9.3.4 Macrocycle and Elementary Cycle

In process industries, continuous control of process variables is a basic requirement for maintaining proper health of the system. A precise cyclic update of process variables is what is needed to ensure the above. Such demands and requirements from process variables are more or less fixed for a given plant. Also occasional and sporadic events (acyclic) such as alarm reporting and set point changes also need to be accommodated in the overall control scheme of things. LAS takes the overall responsibility for such cyclic and acyclic communications in macrocycles. A macrocycle is

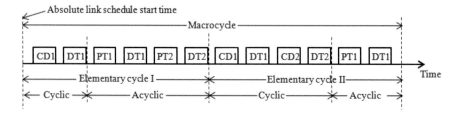

FIGURE 7.16 Macrocycle and elementary cycles.

divided into elementary cycles, which consist of time needed for cyclic and acyclic communications. This is shown in Figure 7.16. It shows a system consisting of two devices requiring cyclic updates. An elementary cycle consists of cyclic and acyclic tasks. Scheduling of periodic tasks (cyclic) is so done that some time is assigned in the elementary cycle to address the need of aperiodic tasks (acyclic), should the need arises.

7.9.3.5 Device Address Assignment

For a device to work properly on a fieldbus, two requirements must be met: the device must have a unique network address and a physical device tag. Network address assignments are done by configuration tool using system management (SM) services.

Assignments of network address follow the following steps:

- Unconfigured (new to the network) devices join the network at one of the four special default addresses.
- A configuration tool chooses a used permanent address and assigns the same to the device using system management services.
- The same configuration tool will then assign a physical device tag to the device using system management services. The above procedure is followed for all new devices that want to join the network. These newly entered network devices store their physical device tag and node address in non-volatile memory. It ensures the devices retain their separate identifications even in case of power failure or system shutdown.

7.9.3.6 Tag Service

A device or a variable can be traced with the help of tag service. System management supports such a service. When a *find tag query* is broadcast to all the connected field devices, each of them starts searching its VFD for the requested tag and returns back all the required information, in case the tag is found, that includes network address, VFD number, VCR index, and OD index. Once the complete path is known, the host or maintenance device can access the data from the tag found.

7.9.4 TRANSDUCER BLOCK

A transducer block models sensors and actuators and connects function blocks to local input/output functions. Transducer blocks isolate the function blocks from the hardware details of a given device. Sensor outputs are accepted by transducer blocks and then written to actuators. Traditional sensors like pressure and temperature transmitters can be mapped into a transducer block. It is connected to a function block through the channel parameter of the function block.

While a particular function block carries out the function that is assigned to it, a transducer block is dependent on the type of measurement. That is for temperature, pressure, or flow measurements, it employs differing measurement principles, but in all the cases it provides an analog value. A transducer block performs various functions like digitizing, scaling, filtering, etc. needed to convert the sensor output to a value which can be accepted by the function block.

Generally there is one transducer block per device channel. Multiplexers allow multiple channels for a single transducer block. The different blocks viz., resource, transducer, and function blocks and their interconnections are shown in Figure 7.17.

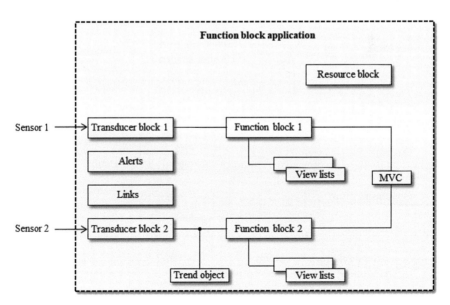

FIGURE 7.17   Different blocks and object representation in user layer. (Courtesy: Technical Overview-Foundation Fieldbus. www.fieldbu.org)

7.9.5 Support Objects

Some additional objects are defined in the user application: link object, trend object, alert object, multivariable container (MVC) object, and view object. Link objects define the links between various function block inputs and outputs internal to the device and across the fieldbus network. Trend objects are used for local trending of function block parameters for host accessing or other devices. Alert objects allow alarm reporting and events on the fieldbus. MVC objects serve to encapsulate multiple function block parameters to optimize communications for report distribution and publishing subscriber transactions. View objects refer to some predefined groupings of block parameter sets that can be displayed by HMI.

7.10 LINKING AND SCHEDULING OF BLOCKS

Figure 7.18 shows how the function blocks in a device or a function block and a VCR are connected via link objects. It shows the PID control of a process consisting of AI, PID, and AO function blocks.

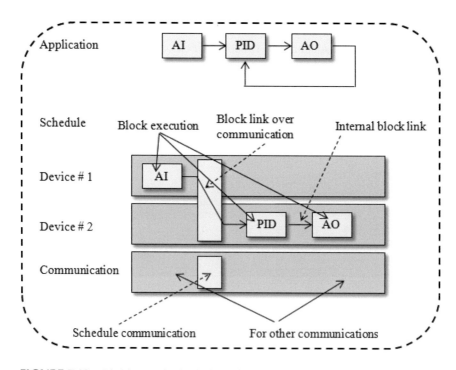

FIGURE 7.18 Linking and scheduling of blocks. (Courtesy: Yokogawa Electric Corporation Foundation Fieldbus Book-A Tutorial, Technical Information TI38K02A01-01E, 2000.)

A function block must get its input parameters before execution of its algorithm. The executed algorithm is then published over the fieldbus. The system management in a field device starts the function blocks according to the function block schedule, which is preprogrammed. The publishing of the executed algorithm is done by LAS which sends a CD schedule to the device. The LAS schedule and the function block schedule are defined as offsets in their respective macrocycles. They must be properly configured so that their actions—execution of actions by the function blocks and publication of data by the LAS on the bus must not overlap.

7.11   DEVICE INFORMATION

Devices used on the fieldbus need to have HMI, fieldbus configuration, and maintenance procedure for their proper operation. These need more information about the devices. Foundation Fieldbus has standardized several files to make devices interoperable with each other.

7.11.1   DEVICE DESCRIPTION (DD)

In Foundation Fieldbus technology, a device is provided with three device support files to ensure interoperability: two device description (DD) files and one capability file. A DD provides information about blocks. A function block parameter can be read and displayed properly by its data type and display specifications. A device, when it joins the system, needs to install its DD to use it with full functionality without any need to update the host software. DDs are platform and system independent.

A DD can be thought of as a *driver* of a device—like the drivers that a PC has to operate printer, scanner, and other devices connected to it. A control system or host can operate with the device only if it has the device's DD.

7.11.2   DEVICE DESCRIPTION LANGUAGE (DDL)

DDs are written in a standardized programming language called device description language (DDL). Device functionality and data semantics can be described by DDL. This is then compiled with *tokenizer* software to generate *DD binary* files.

A DD binary has two files: a DD binary with extension *.ffo* and DD symbol list with extension *.sym*. Once these two files are installed, full access to the device would be ensured. Instead of writing the full functionalities of each device individually, the common portion of the devices' functionalities is included in DD library, which includes a dictionary also. The special features of each device are included in DDL.

7.11.3 DD Tokenizer

It is a software tool which converts DD source input files into DD output files by replacing key words and standard strings in the source file with fixed tokens. Standard features associated with resource, function, and transducer blocks are included in DD library with the help of tokenizers.

7.11.4 DD Services (DDS)

Device description services (DDS) is a software for human machine interface. It is a library function which helps in reading the device descriptions. A device can be added to the system and it would work nicely if the DD of this device is added to the host or control system.

With the help of DDS, device descriptions are made available only to the system and not their physical values. The latter can be read from the FMS communication services. DDS helps in connecting devices from different manufacturers on the same fieldbus system.

7.11.5 DD Hierarchy

The hierarchical structure of DD is shown in Figure 7.19. Fieldbus Foundation has defined such a structure so that configuration of devices becomes very easy. There are four levels in the hierarchy and are referred to as: universal parameters, function block parameters, transducer block parameters, and manufacturer specific parameters.

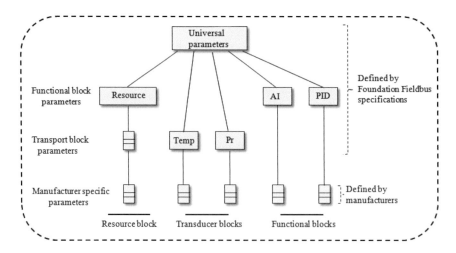

FIGURE 7.19 Hierarchical structure of DD. (Courtesy: SMAR, Fieldbus Tutorial-A Foundation Fieldbus Technology Overview, USA, Available at http://www.smar.com/PDFs/catalogues/FBTUTCE.pdf.)

Universal parameters refer to common attributes like revision, mode, tag, etc. All blocks must include universal parameters. Parameters for standard function blocks are defined in function block parameters. In transducer function blocks, parameters for processes are defined like pressure, flow, temperature, etc. In the manufacturer specific parameters, a manufacturer may add additional parameters to the function block parameters and transducer block parameters.

7.11.6 CAPABILITIES FILE

Capabilities file tells the host about the resources a device has in terms of function blocks and VCRs. This enables the host to configure a device offline. It is the responsibility of the host to ensure that only functions supported by the device are allocated to it. A capabilities file has an extension *.cff*. Capabilities file is often called *CFF*, which stands for common file format.

7.11.7 DEVICE IDENTIFICATION

A device on the fieldbus can be identified in one of the three following ways: device identifier (ID), physical device (PD) tag, and node (physical) address.

A device ID is a 32 byte unique number and no two devices can have the same device ID. The device ID is burnt into the device by the manufacturer of the device and can never be changed. The PD Tag is a unique name assigned to the device in the plant by the user. It is used to identify a device for the specific purpose (use) in the plant. It is also 32 bytes in length. When a device is replaced, the new device is assigned the old device's PD Tag number. The physical or node address is a unique 1 byte number in a fieldbus segment and is assigned by the user at the time of configuring the network. Since the node address length is very small compared to the other two, it is commonly used when communication is needed in the network.

7.12   REDUNDANCY

Redundancy in a fieldbus system is included at different levels to ensure availability of resources in times of breakdown. Redundancy can be added at host level, device level, media level, and network level. HSE has built-in robustness to ensure fault tolerance. Decentralization of operations is another measure of fault tolerance. When many controls are included in a single controller, failure of the same would mean a considerable part of the plant to be out of order and it may have some catastrophic influence on the safety of the plant. Decentralization of operations of the loops would ensure a small part of the plant to be out of order in case of single controller failure. Another major advantage of decentralization is localizing the fault.

7.12.1 HOST LEVEL REDUNDANCY

Redundancy at the host or central level is used to ensure high availability and thereby prevent disruption in production and consequent downtime and heavy losses. Host level ties the network and the field level devices together and any breakdown here may be life threatening for the entire plant. There can be various methods to achieve host-level redundancy: media redundancy, network redundancy, and network and media redundancy.

7.12.1.1 Media Redundancy

Media redundancy is independent of the protocol used. Figure 7.20 shows a media redundancy scheme that uses star topology having redundant port receivers (dual port transreceivers). Such a transreceiver has three ports: one for the device itself and the rest two are for primary and secondary communication paths. The transreceiver switches from the primary to the secondary path when the former loses its link. When the transreceiver fails, both the links are lost and consequently both the paths fail.

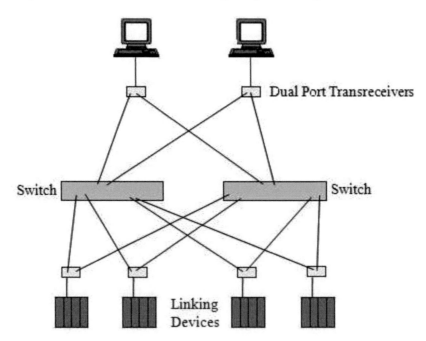

FIGURE 7.20 Media redundancy using dual port transreceivers. (Courtesy: J. Berge. Fieldbuses for Process Control: Engineering, Operation and Maintenance, ISA, p. 143, USA, 2004.)

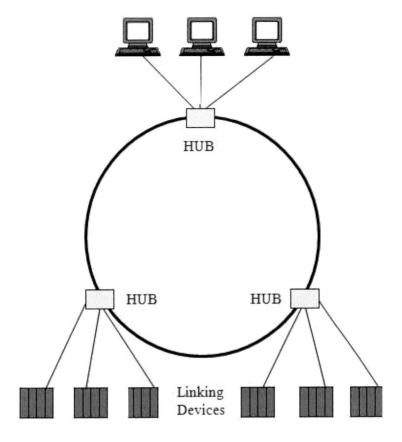

FIGURE 7.21 Media redundancy using single ring topology. (Courtesy: J. Berge. Fieldbuses for Process Control: Engineering, Operation and Maintenance, ISA, p. 144, USA, 2004.)

Figure 7.21 uses a single ring topology to achieve media redundancy. The linking devices are connected to the hubs. The topology provides a dual communication path—both clockwise and anticlockwise. When one path fails, communication can still take place through the other.

It is advisable to use several smaller hubs instead of a central hub connecting many devices. Failure of a smaller hub impacts the system in a lesser way than the failure of a bigger hub would on the communication. Again the distance from a hub to a device should be kept small to reduce the chances of damaging a non-redundant cable segment.

7.12.1.2 Network Redundancy

A complete host-level network redundancy scheme is shown in Figure 7.22. For its implementation, it needs to have two separate communication paths at each stage. It is used when very high availability is desired. The primary

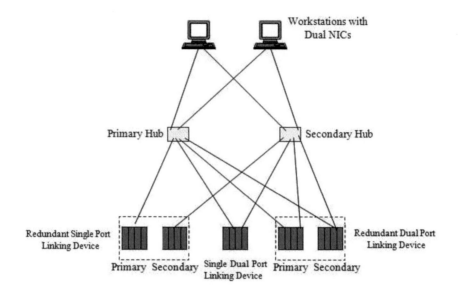

FIGURE 7.22 A network redundancy scheme. (Courtesy: J. Berge. Fieldbuses for Process Control: Engineering, Operation and Maintenance, ISA, p. 20, USA, 2004.)

and its corresponding secondary (redundant) device must be kept on separate networks.

The scheme can be implemented in two ways; the primary and secondary devices of the redundant device may have a single port. Alternatively, the primary and secondary devices may have two ports. Both the ports have self diagnostics, whereas communication takes place via a port at any given time. In this scheme, workstations are provided with dual network interface cards (NICs).

Alternate port and segment are automatically used in case of failure of primary network path. Such a changeover is totally bump less and transparent to the operator. Each and every port on the network has its own separate IP address, making diagnostics very easy.

7.12.2 SENSOR REDUNDANCY

Sensor redundancy is achieved by employing at least two sensors to measure the value of a process variable at a single point. The two sensor outputs are connected to the two input points of the transmitter. With the help of the standard selector block in the Foundation Fieldbus programming language, the better of the two sensor outputs is selected. In case of failure of the *good* sensor, the other sensor takes over, thereby obviating the need of shutdown. Thus, redundant sensors result in less number of shutdowns.

Only if both the sensors fail, it would result in shutdown. This is an example of two out of two (2oo2) configuration.

7.12.3 TRANSMITTER REDUNDANCY

Transmitter redundancy involves employing several transmitters for accessing the same process point. Foundation Fieldbus uses status bytes for each process variable to select the *good* transmitter output and reject the others. Foundation Fieldbus takes the help of standard selector block for the above. Such a selector block can handle three or even more inputs. Initially the configuration strategy is so designed that in case of failure of a transmitter, the other automatically takes over. This eliminates any possibility of the loop shutdown of which these transmitters are a part.

When four such transmitters are so used for accessing and connecting the same process point, this is called four out of four (4oo4). Transmitters from different manufacturers are used to eliminate identical type of transmitter failure and thus increasing transmitter redundancy. Foundation Fieldbus technology ensures that transmitters from different manufacturers are interoperable.

7.13   HSE DEVICE TYPES

HSE devices are of four types. These are: host device, linking device, gateway device and Ethernet device. A combination of these four device types are sometimes used to form a single device type.

Host devices are non-HSE devices which can communicate with HSE devices. Examples include operator workstations, open platform communication (OPC) server, and configurators. A linking device is one which connects H1 networks to the HSE network. It provides gateway services that map FDA, SM, and FMS messages to their counterparts. A gateway device interfaces other network protocols (for example, Profibus, Modbus, etc.) so that they can work seamlessly with Foundation Fieldbus protocols. The Ethernet device provides services for communication over Ethernet media, Ethernet stack initialization, and time synchronization. Function blocks may be executed by an Ethernet device.

7.14   SYSTEM CONFIGURATION

There are two phases for configuring a system, which are system design and device configuration. The system design has to be completed first as per the requirements of the system under consideration and then only the device configuration is done. The system becomes operational only after the devices have been configured in the right manner.

7.14.1 SYSTEM DESIGN

For Foundation Fieldbus systems (or for that matter for any fieldbus system), the system design is akin to distributed control system (DCS). But a fieldbus system design differs from a DCS based design on two counts. First, the conventional point-to-point 4-20 mA current loop is replaced by a digital bus where many smart field devices are connected to the same bus. Second, the control and I/O functions are brought into the fold of fieldbus devices. The ability of fieldbus devices to distribute the functions in these devices drastically reduces the number of remote mounted I/O equipment as also the number of rack mounted controllers. The devices residing on the fieldbus must have their individual unique physical device tag and a corresponding network address.

7.14.2 DEVICE CONFIGURATION

Following control strategy, devices are configured by connecting the function block inputs and outputs together for each device. Once the above are completed, each such fieldbus device would generate its own information. The system becomes operational once all the fieldbus devices are configured and their individual input and output blocks are interconnected as per the control strategy.

8 PROFIBUS

8.1 INTRODUCTION

PROFIBUS (PROcess FIeldBUS) is an open fieldbus standard developed in 1989 catering to the needs of both process automation and factory manufacturing automation. Initially developed by Siemens, it is particularly suitable for fast, time critical applications and also involving complex communications. It is an open, vendor-independent (thus providing interoperability) fieldbus standard which adheres to open systems interconnection (OSI)/International Standards Organization (ISO) model for communication. It supports single cable wiring having multi-input sensor blocks, intelligent devices, operator interfaces, and smaller sub-networks like AS-i. It is based on German National Standard DIN 19 245 Parts 1 and 2 and also has been ratified by the European National Standard EN 50170 Vol. 2. The general PROFIBUS features are shown in Table 8.1.

PROFIBUS supports two types of devices: master device and slave device. The former is called an *active station* while the latter is called a *passive station*. A master device has the right to control the bus when it has bus access. Then it can transmit messages without any remote request. Transmitters, sensors, and actuators are examples of slave devices. A slave device acknowledges any received message and on receiving a request from a master, can send messages to that master.

8.2 THE PROFIBUS FAMILY

Several PROFIBUS standards are there which are: PROFIBUS DP (master/slave), PROFIBUS FMS (multimaster/peer-to-peer), and PROFIBUS PA (intrinsically safe). DP, FMS, and PA stand for **D**ecentralized **P**heriphery, **F**ieldbus **M**essage **S**pecification, and **P**rocess **A**utomation, respectively.

PROFIBUS DP handles fast communication processes like drives, remote input/outputs (I/Os) normally encountered in factory automation. In this mode, multimasters are also used in which case a slave is assigned to one master only. It means that multiple masters can read inputs from a specific device but only one master can write outputs to that device.

Field devices are generally connected to PROFIBUS-PA. It differs with PROFIBUS-DP on three major counts, which are: the devices can be powered on the bus cable, it supports devices in explosion hazardous areas,

TABLE 8.1
Salient PROFIBUS Features

Communication methods	Master-slave, multimaster slave Publisher-subscriber	
Network speed	9.6 kbps–12 Mbps	
Data transfer rate	Up to 244 bytes	
Transmission Media/ technologies	RS 485 STP copper: 126 Fiber optic: 126 IR: 126	
Max. No. nodes	RF: 126 Slip ring: 126 MBP-IS: depends on power budget	
Max. Distance	RS 485 STP copper segment/ with 9 repeaters	9.6 kbps: 1000 m/10000 m 12 Mbps: 100 m/1000 m
	Fiber optic (between fiber optic repeaters)	Plastic: 50 m Multimode glass: 400 m Single mode glass: 15 km
	IR & RF	Varies with vendor product
	MBP-IS	31.25 kbps: 1.9 km max depending on cable type
Diagnostics	Standard: 6 bytes Detailed: Up to 238 bytes total	Device related Module related Channel related

data transfers on the physical layer takes place vide IEC 61158-2. It allows higher freedom in the selection of bus topology and longer bus segments.

PROFIBUS FMS is a peer-to-peer messaging format. This enables masters to communicate with each other. Up to 126 nodes are available, like PROFIBUS DP and all can be masters. FMS messages have more overhead than PROFIBUS DP messages.

Apart from the above three types, a *combi mode* is sometimes used which uses both FMS and DP simultaneously in the same network. This is used when a PLC is used along with a PC, in which case the primary master communicates with the secondary master via FMS.

8.3 TRANSMISSION TECHNOLOGY

Different transmission technologies are used in PROFIBUS which can be RS485, RS485-IS, MBP, and Fiber Optic. Table 8.2 summarizes different transmission technologies with regard to their data rate, maximum cable

TABLE 8.2

Physical Layer Transmission Technology

	MBP	RS485	RS485-IS	Fiber Optic
Data transmission	Digital, bit synchronous, Manchester encoding	Digital, differential Signals according to RS485, NRZ	Digital, differential signals according to RS485, NRZ	Optical, digital, NRZ
Transmission rate	31.25 kbps	9.6–12,000 kbps	9.6–1,500 kbps	9.6–12,000 kbps
Data security	Preamble, error protected, start/end delimiter	HD = 4, Parity bit, start/end delimiter	HD = 4, Parity bit, start/end delimiter	HD = 4, Parity bit, start/end delimiter
Cable	Shielded, twisted Pair, copper	Shielded, twisted pair copper, cable type A	Shielded, twisted 4-wire, cable type A	Multimode glass fiber, single mode glass fiber, PCF, plastic
Remote feeding	Optional available over single wire	Available over additional wire	Available over additional wire	Available over hybrid line
Protection type	Intrinsic safety (EEx ia/ib)	None	Intrinsic safety (EEx ib)	None
Topology	Line and tree topology with termination; also in combination	Line topology with termination	Line topology with termination	Star and ring Topology typical; line topology possible
Number of Stations	Up to 32 stations per segment; total sum of max. 126 per network	Up to 32 stations per segment without repeater; up to 126 stations with repeater	Up to 32 stations per segment, up to 126 stations with repeater	Up to 126 stations per network
Number of repeaters	Maximum 4 repeaters	Maximum 9 repeaters with signal refreshing	Maximum 9 repeaters with signal refreshing	Unlimited with signal refreshing (time Delay of signal)

Source: www.pacontrol.com/download/profibus-overview.pdf

length, protection, cable types, safety issues, data security, topology, etc. at the physical level.

The most commonly used is RS485 which uses a shielded twisted pair cable with transmission rates up to 12 Mbps. The bus structure used allows the addition and deletion of a station without affecting other stations. In a particular segment, there can be up to 32 devices, including master and slave devices. Either side of a segment is terminated with an active bus terminator. When more than 32 devices are used or there is a need to expand the network, repeaters are used.

RS485-IS is used in potentially explosive areas (type EEx-i) that uses a 4-wire medium. When using this technology, maximum current and voltage levels must be adhered to avoid any potential explosive situation.

MBP stands for Manchester coding with bus powered. It is a synchronous transmission scheme with 31.25 kbps rate of transmission. It is mainly used for process automation systems with special emphasis on chemical and petrochemical industries.

Industries having high electromagnetic disturbances and devices spaced at considerable distances employ fiber optic transmission schemes. Obviously, devices in the network must be able to integrate with the fiber optic transmission technology in the physical layer.

8.4 COMMUNICATION PROTOCOLS

The three PROFIBUS versions, namely FMS, DP, and PA, all use a standard bus access protocol. This protocol is implemented by layer 2 of OSI which in the case of PROFIBUS, is termed as fieldbus data link (FDL). Along with handling transmission protocols, FDL handles both data security and error detection.

The protocol is so designed that the three variants work seamlessly together by offering high speed, high deterministic operation at the field level, reduced costing by employing two wire connections for PA, and an extended capability at the control level for FMS.

Of the three protocols, FMS is the first communications protocol and is designed for operations at the cell level, where normally PLCs and PCs communicate with each other. A wide range of functions are offered by FMS making it quite complex to implement at an average transmission speed.

DP has three variants, namely DPV0 to DPV2. The original one, i.e., DPV0 offers basic functionalities which includes cyclic I/O communication and diagnostic reporting. DPV1 offers both cyclic and acyclic communication and alarm services, diagnostics, parameterization and field device controls, while DPV2 supports some services which are particularly needed in the field of drive control. These include functions for producer-consumer communication between slave devices, isochronous slave mode, time synchronization, and time stamping. PROFIBUS DP can address extremely time-critical communication tasks. The last one—PROFIBUS PA is specially oriented toward meeting the requirements of process automation communication.

Layer 2 (the data link layer (DLL) or FDL as it is termed here) corresponds to the bus access protocol which manages the communication procedure between the master-slave and the token passing method for multimaster system. FDL also handles data security and data frames. Layer 7 or the application layer acts as an interface between the application programs and the different profiles—FMS, DP, or PA existing in the user layer.

8.5   DEVICE CLASSES

PROFIBUS devices can be classified into three device types: Class 1 PROFIBUS DP Master (DPM1), Class 2 PROFIBUS DP Master (DPM2), and PROFIBUS slaves.

Class 1 masters are usually used for cyclic data exchange with slaves connected to it. They are normally PLCs or PCs programmed for data exchange with slaves on a precise time sharing basis. Class 1 masters have the following characteristics: tokens are passed between masters, can write data into the slaves assigned to it and can read data from a slave in the network, sets the data rate and the connected slaves detect the same automatically.

Class 2 masters are used as a tool for device and system commissioning. In DPV1 and DPV2, this master class is used for setting/altering device parameter values acyclically. DPM2 are not required to remain permanently connected to the system and are used only for initial device configuration. Normally, PROFIBUS master devices support the functionalities of both DPM1 and DPM2. Class 2 masters have the following characteristics: act as supervisory masters, used for diagnostic purposes and slave commissioning, control slaves at any given point of time, can only read slaves but do not have write access.

PROFIBUS slave devices respond to master polling by sending device data. The slaves are field devices like transducers, valves, remote I/Os, etc. Slaves can be of two types: compact devices and modular devices. Compact devices have a fixed I/O configuration, unlike the case of a modular device in which it may vary. A modular device contains a station to which the fieldbus interfaces are connected through individual slots specifically meant for them. Multivariable devices can be thought of as modular devices.

8.6   PROFIBUS IN AUTOMATION

Figure 8.1 shows how PROFIBUS technology is applied at different levels in the automation hierarchy. The three different versions of PROFIBUS, namely DP, PA, and FMS, all share the same transmission medium RS-485. The figure shows a typical network having both process automation and factory automation tasks in the same network.

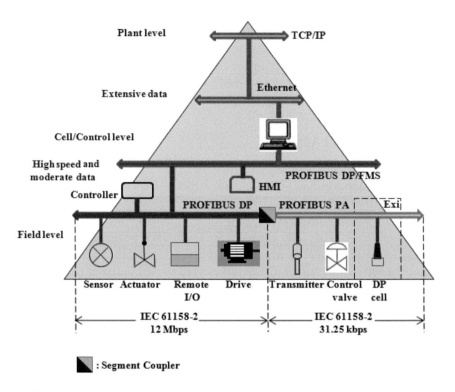

FIGUE 8.1 A typical PROFIBUS network includes both process and factory automation. (Courtesy: A. A. Verwer. Introduction to PROFIBUS. Manchester Metropolitan University, Department of Engineering, Automation Systems Center, PROFIBUS Competence Center, 2005.)

8.7 OSI MODEL OF PROFIBUS PROTOCOL STACK

A uniform bus access mechanism is followed by all the three versions (DP, PA, and FMS) of a PROFIBUS system. This protocol is implemented by layer2 of the OSI model. This layer in PROFIBUS is known as FDL. The PROFIBUS protocol stack is shown in Figure 8.2.

The PROFIBUS protocol ensures that all the three variants work seamlessly and provides high speed, deterministic operation and communication at the field level (levels 1 and 2), DP (level 7), and user/cell level (FMS).

8.8 PROFIBUS DP CHARACTERISTICS

PROFIBUS is used for fast communications at the device level of process automation systems like chemical, paper, food, automobile industries, etc. It is included in EN 50254 and IEC 61158 standards.

PROFIBUS DP is the high-speed solution for PROFIBUS systems. It is largely used in time critical solutions. Originally it was developed for communication between automation systems and decentralized equipment.

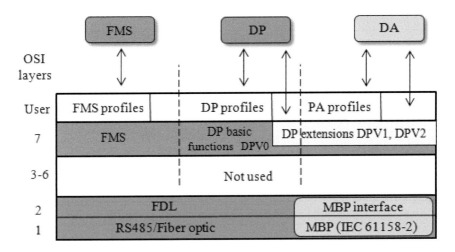

FIGURE 8.2 PROFIBUS protocol stack related to ISO/OSI model. (Courtesy: A. A. Verwer. Introduction to PROFIBUS. Manchester Metropolitan University, Department of Engineering, Automation Systems Center, PROFIBUS Competence Center, 2005.)

Figure 8.3 shows the bus cycle times for a DP mono master system assuming that each slave has two bytes of input and output data. From the figure, it is seen that approximately 1 ms time would be required at a bit rate of 12 Mbits/sec for transmission of 512 bits of input and 512 bits of output data distributed over 32 stations. Bus cycle time would depend on both the number of stations and the transmission rate.

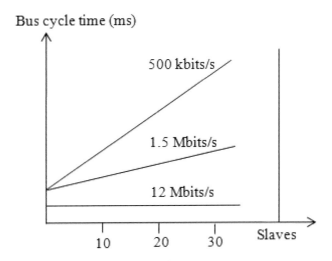

FIGURE 8.3 Bus cycle time for a DP mono master system. (Courtesy: PROFIBUS, Technology and Application: System Description, PROFIBUS International Support Center, Karlsruhe, Germany, Copyright by PNO 10/02, 2002.)

8.8.1 VERSION DP V0

Basic functionalities of DP are provided in DPV0. These include cyclic data exchange between master and slaves, module diagnosis, channel-specific diagnosis, station diagnosis, etc.

The master reads the input information from the slaves in a cyclic manner and writes output information into the respective slaves cyclically.

8.8.1.1 Diagnostic Functions

Diagnostic functions are used for fast location of faults. The diagnostic message is sent over the bus. Such functions can be of three types: device-specific diagnosis, module-related diagnosis, and channel-related diagnosis.

Any device-specific parameter like, over voltage, under voltage or over-heating is taken care of by device-specific diagnosis services. Module-related diagnosis relates to any pending diagnosis in a specific I/O sub domain of a station. Any fault related to an individual input/output bit (channel) is manifested in channel-related diagnosis.

8.8.1.2 Synchronization and Freeze Mode

DPM1 handles cyclic data transfers automatically. In addition to this, it can send control commands to a slave or a group of slaves simultaneously. These are called multicast commands. A master can send a *sync* command to a slave or a group of slaves. The outputs of the addressed slaves are then frozen in their current states. During subsequent data transmissions, the output data are stored in the slave and the output state remains unaltered. The stored output data are not sent to the outputs until the next sync command is received by the slave. A sync command can be terminated by issuing an *unsync* command.

When a master issues a *freeze* command to a slave, the corresponding slave enters into the freeze mode. Then the states of the slave inputs are frozen at their current values. This remains so until the slave receives another freeze command, when the input data is updated.

8.8.1.3 System Configuration

PROFIBUS systems support mono master and multimaster systems. A maximum of 126 devices, which includes both master and slave, can be connected to the bus. System configuration involves the following: number of stations, bus parameters, diagnosis message formats, assigning station addresses to the I/Os.

8.8.1.4 Time Monitors

A time monitor is a very efficient and effective protective mechanism to combat incorrect parameterization or failure of transmission-related functions. Time monitoring mechanisms are fitted both at the master and the slaves. During configuration, the time interval of a time monitor is specified.

At the master level, DPM1 uses a data_control_timer function to monitor data communication with the slaves. Each slave has its own time monitor. The timer is tripped if no correct user data transfer takes place within the specified time interval and the user is notified accordingly.

Slaves use their respective watchdog timers to detect errors of the master or transmission error. The outputs of the slaves are put into fail safe mode if no data communication with the master occurs within the watchdog control interval, which is preset.

In a multimaster system, it must be ensured that only the authorized master gets the access to the corresponding slave.

8.8.1.5   Token Passing Characteristics

The characteristics associated with token passing mechanism are discussed as follows:

- Token passing technique involves more than one master.
- The token is passed from one master to another in ascending order.
- Each master is responsible for the addition or removal of stations in its address range.
- The system does not need initialization in case of loss of the token. The master with the lowest address creates a new token after its token timer is timed out.
- After a power up, the master station with the lowest station address, commences initialization. It claims the token if it sees no bus activity after waiting for a predefined period. It informs other master stations on the network about its own activity and transmits a *request field data link status* to each station in an increasing order. The station which responds to this request is passed on the token.

8.8.2   VERSION DPV1

DPV1 is an enhancement over DPV0 in process automation and control field. Both cyclic and acyclic communications are implemented in this version. This permits online access to stations using engineering tools. In addition, parameter assignment, alarm handling of intelligent devices, calibration of field devices, etc. are also handled by this version.

8.8.2.1   Cyclic and Acyclic Communication

Both cyclic and acyclic communications are features of PROFIBUS DPV2. DPM1 devices like PLCs exchange information with slaves cyclically at predefined time slots. The master has the token and sends messages to or

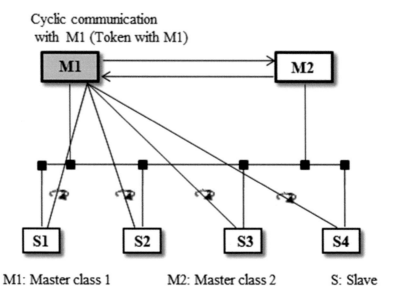

FIGURE 8.4 PROFIBUS DP bus access by cyclic communication. (Courtesy: B. G. Liptak. Instrument Engineers' Handbook, Process Software Digital Network, 3rd Edition, CRC Press, Boca Raton, FL, p. 580, 2002)

retrieves them from slave 1 and then from slave 2 etc. in that sequence until it reaches the last slave in the list. The master can then utilize the remaining available time of the program cycle to set up an acyclic communication with any slave. The cyclic and acyclic communication services are shown in Figures 8.4 and 8.5, respectively.

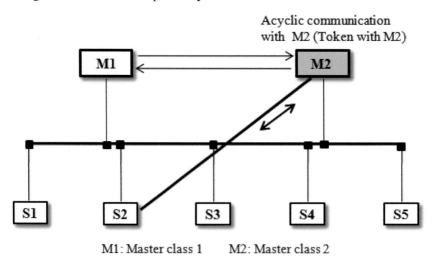

FIGURE 8.5 PROFIBUS DP bus access by acyclic communication. (Courtesy: B. G. Liptak. Instrument Engineers' Handbook, Process Software Digital Network, 3rd Edition, CRC Press, Boca Raton, FL, p. 580, 2002)

Cyclic communication is mostly carried out in a transparent manner. The master has its own memory in which it stores the incoming data from the slaves. Any data that is coming out from the master is first written into the master's memory before being sent out to the concerned slave.

8.8.3   Version DPV2

Version DPV2 is still another improvement over DPV1. Additional functionalities of DPV2 include isochronous mode, direct slave-to-slave communication without any help from the coordinating master. This publisher–subscriber model helps reduce bus response times by up to 90%. DPV2 can also be implemented as a drive bus to control fast movement sequences in drive axes.

8.8.3.1   Slave-to-Slave Communication

It is a direct communication between slave-to-slave without going through the master. It saves time considerably, to the extent of up to 90 percent. One slave acts as publisher and the other ones as subscribers. This occurs without the master. Thus, this process enables a slave(s) (subscriber(s)) to read data from another slave (publisher). Figure 8.6 explains the above process.

8.8.3.2   Isochronous Mode

This facilitates clock synchronization between masters and slaves irrespective of bus load. It provides highly precise positioning systems with clocks

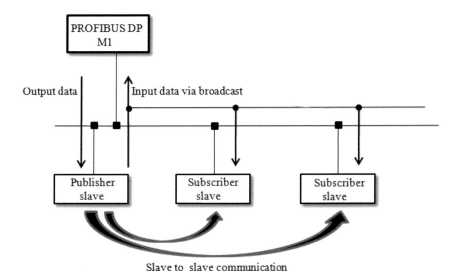

FIGURE 8.6   Slave to slave communication

which are precise within one microsecond. All the devices connected to the network are synchronized with the master through a *global control broadcast message*. A *special sign of life* enables monitoring of synchronization.

8.8.3.3 Clock Control

All slaves connected to a master are synchronized by *clock control*. That is, all slaves are time stamped by the master to within less than a millisecond. This time stamping of the slaves are done by the new connectionless MS3 channel. Clock control is a major facility which allows the network to track occurrence of events very precisely. This is particularly useful for a multimaster system. Clock control helps in diagnosing and identification of faults as well as chronological planning of events.

8.8.3.4 Upload and Download

This is a special function which enables loading of data of any size in a device with very few commands. It allows programs to be updated or devices replaced without any need for manual loading processes.

8.8.3.5 HART on DP

A large installation base of HART devices has led many users to integrate PROFIBUS systems on them. An immediate advantage of the HART devices is that they can benefit from the PROFIBUS communication techniques.

For such integration of HART on DP, PROFIBUS has implemented a specification profile above layer 7 in both master and slave devices. This ensures the mapping of HART client-master-server model on the PROFIBUS. Figure 8.7 shows such an integration in which the

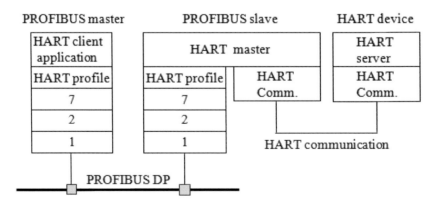

FIGURE 8.7 Integrating HART into PROFIBUS DP. (Courtesy: PROFIBUS, Technology and Application: System Description, PROFIBUS International Support Center, Karlsruhe, Germany, Copyright by PNO 10/02, 2002.)

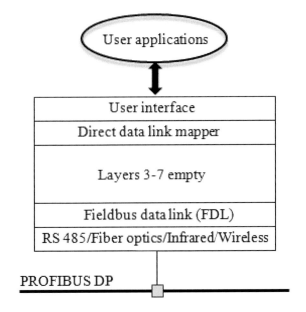

FIGURE 8.8 Communication profile of PROFIBUS DP. (Courtesy: B. G. Liptak. Instrument Engineers' Handbook, Process Software Digital Network, 3rd Edition, CRC Press, Boca Raton, FL, p. 623, 2002)

HART-client application is integrated in a PROFIBUS master and the HART master in a PROFIBUS slave.

8.8.4 COMMUNICATION PROFILE

The communication profile of PROFIBUS DP is shown in Figure 8.8. It shows the physical layer which uses different physical mediums such as fiber optics, infrared, or other wireless communication systems. Layer 2, the data link layer, is called FDL here. Layers 3 to 7 are empty in PROFIBUS DP configurations. Other functions required in a fieldbus are implemented by two upper layers called direct data link mapper (DDLM) and user interface.

8.8.5 PHYSICAL LAYER

Mostly, the physical layer of PROFIBUS DP is specified by R-S485 standard. Some other configurations involve using fiber optics, infrared, or other wireless transmission medium. For RS-485, the physical medium used is a twisted pair cable which may be shielded, if required. A maximum of 126 stations can be connected.

TABLE 8.3

Transmission Speed vs. Maximum Segment Length for PROFIBUS DP

Transmission speed (kbps)	9.6	19.2	93.75	187.5	500	1500	6000	12000
Maximum segment length (m)	1200	1200	1200	1000	400	200	100	100

The physical layer of PROFIBUS DP, based on RS-485, has the following features:

- The topology is a linear bus and terminated at both ends.
- Transmission is via a twisted pair cable, with optional shielding. For transmission rates greater than 500 kbaud, Type A cable is preferred, while Type B cable is used for short distance communication and having a baud rate lesser than the above.
- Stubs are possible along the line.
- Different physical media may be used like; fiber optic, infrared, wireless transmission etc.
- PROFIBUS DP, FMS and PA can together be used on a common bus line.

8.8.5.1   Transmission Speed vs. Segment Lengths

Transmission speed varies between 9.6 kbps and 12 Mbps and the maximum length of a segment is related to transmission speed, as shown in Table 8.3.

8.8.6   DATA LINK LAYER

Data link layer in PROFIBUS DP is known as FDL. It performs the following tasks:

- Bus access control or medium access control—MAC
- Data security
- Telegram structure
- Availability of data transmission services
  - o SRD (send and request data with reply) and
  - o SDN (serial data with no acknowledge)

Data communication between master and slave takes place via the logical token passing method. Every station knows the address of the station from which it received the token and also the address of the station to which the token is to be handed over.

Data transmission with SRD is a confirmed one and the source station can send a maximum of 246 bytes to the selected destination station. The latter responds with a maximum of 246 response bytes. With SRD, the master either issues a command or sends data to the concerned slave and receives a reply (either an acknowledgement or data from the slave) in a defined time span. For SDN services, no confirmation is forthcoming from the destination station. The source station can send a maximum of 246 bytes—as is the case with SRD. SDN is mainly used for broadcast or multicast messages.

8.9   PROFIBUS PA CHARACTERISTICS

PROFIBUS PA caters to the needs of process automation and control systems. The field instruments, viz., transmitters, positioners, valve actuators, etc. are all connected by a PROFIBUS PA segment. The connection to PROFIBUS DP is made by using a DP/PA coupler.

Potential advantages accrue by using the PA technology such as failure safety systems, auto diagnosis, reliable information transfer, equipment rangeability, high-resolution measurement, integration to higher speed discrete control, etc. It results in significant cost reduction in installation, less downtime, higher reliability of operations, easy integrability for future expansion, more functionality and safety, less start up time, etc.

PROFIBUS PA communication uses identical protocol with that of PROFIBUS DP. The twisted pair cable is used for both data communication as well as power supply to the devices. However, three major differences exist with PROFIBUS DP which are given as follows:

- devices are powered on the bus cable
- devices can be used in explosion hazardous areas and
- data are transmitted via the IEC 61158-2 physical layer

In PROFIBUS PA, the IEC transmission specifications define a digital bit synchronous data transmission at a rate of 31.25 kbits/sec. The topology can be line or tree with 126 addressable devices and up to 1900 meter line length.

8.9.1   BUS ACCESS METHOD

Master-slave configuration is used in PROFIBUS PA systems to regulate bus access. In a multimaster system, token passing method is used so that at any instant, only one master is active and regulates the traffic in DP-PA environment. Each master has bus control under it for a precisely defined time. The devices on the PROFIBUS PA segment are accessed by a segment coupler or link. Figure 8.9 shows the diagram of how the PROFIBUS PA accesses the bus via a segment coupler.

Segment couplers are transparent to the PROFIBUS DP master and as such are not engineered in PLC. The couplers only route the signals to the proper devices in the PROFIBUS PA environment. The couplers are not assigned any PROFIBUS DP address. Instead each field device is given a

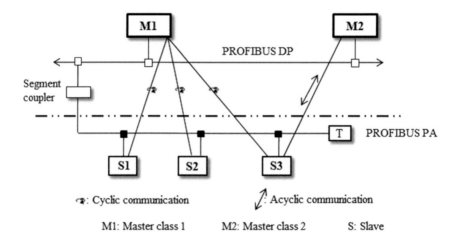

FIGURE 8.9   Bus access in PROFIBUS PA system via segment coupler. (Courtesy: B. G. Liptak. Instrument Engineers' Handbook, Process Software Digital Network, 3rd Edition, CRC Press, Boca Raton, FL, p. 581, 2002)

PROFIBUS DP address through which they communicate with the masters. As far as PROFIBUS DP is concerned, the devices merely act as slaves.

A link, on the other hand, is recognized by the PROFIBUS DP master and must be engineered in PLC. Since the link is opaque in nature, PROFIBUS DP master cannot see the PROFIBUS PA devices. The link acts as a bus master on the PROFIBUS PA side while it acts as a slave on the PROFIBUS DP side. Figure 8.10 shows how the PROFIBUS PA bus access is effected via a link.

Each field device is allocated a PROFIBUS PA address. This address is unique for that link only but is not valid for other PROFIBUS PA segments. The master polls the devices cyclically and stores the respective data values in a buffer.

On the PROFIBUS DP side, the link which acts as a slave is assigned a PROFIBUS DP address. The master polls the devices cyclically and each device data is packed in respective telegrams by the link.

8.9.2   DEVICE PROFILE

A process control system is best accepted by end users if operation, communication and monitoring of device parameters and functions are totally standardized such that interoperability between devices from different manufacturers is ensured. Device profiles ensure that the properties and functions of field devices are predefined within suitable limits. Device profile includes parameterization of different variables like measured value, alarm limits, status flag, scaling factor, etc.

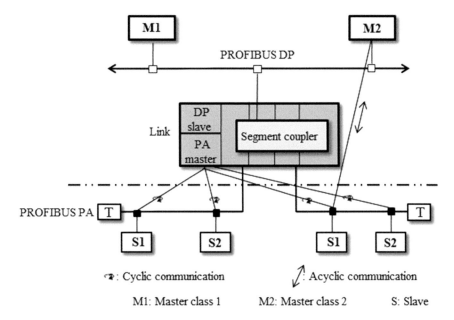

FIGURE 8.10 Bus access in PROFIBUS PA system via a link. (Courtesy: B. G. Liptak. Instrument Engineers' Handbook, Process Software Digital Network, 3rd Edition, CRC Press, Boca Raton, FL, p. 582, 2002)

There are two different profile classes for a PROFIBUS PA system: class A and class B. Class A profiles include basic parameters which are absolutely necessary for process automation systems. These include process variable, status of the measured value, physical unit, and the tag number.

Class B profiles include all the basic parameters of class A. In addition, class B includes several other functions. It differentiates between parameters which are mandatory and those which are optional in nature.

8.9.3 PA Block Model

Blocks are commonly used to describe the characteristics and functions of a measuring point. Automation applications are generally represented through a combination of such several blocks. PROFIBUS PA devices, requiring PNO certification, must have a set of universal parameters which could be implemented. It would then ensure configuration compatibility between identical device types made from different manufacturers.

Figure 8.11 shows a PROFIBUS-PA block model. It presents a complete overview of the PA device in block form and shows the data flow path through the three data channels, viz., MS0, MS1 and MS2. It consists of four blocks: transducer block, physical block, function block, and device

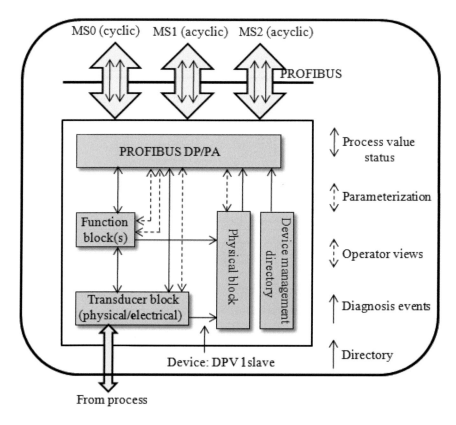

FIGURE 8.11   Block model of a PROFIBUS PA device.

management block. The sensor output first goes to the transducer block from which it goes to the function block. The physical block is not a part of process signal flow but contains information about the device itself, like the serial number of the device, manufacturer code, installation date, and diagnostic features.

Except the device management block, the beginning of every other block contains the standard parameters which are used and identify the block. The parameters assigned to an individual block use the data structure and data formats as used in PROFIBUS standard. The data structure ensures that data stored and transmitted all follow an orderly manner.

8.9.3.1   Transducer Block

The transducer block accepts the sensor output. It processes the same and its output goes to the function block. The measurement principle of the transducer block is dependent on the basic process parameter that is to be measured and controlled. Currently available transducer blocks are: temperature, pressure, level, flow, discrete input/output for switches, electromagnetic and pneumatic transducer blocks for actuators. Multifunctional sensors having more than one sensor will have corresponding number of transducer blocks.

8.9.3.2 Physical Block

It contains the properties of the field device, i.e., device parameters and functions which are independent of the measurement method. It contains the serial number of the device by which it can be uniquely identified, the code of the particular manufacturer, data about the installation date, device operational data and diagnostic features—both standard and manufacturer specific.

8.9.3.3 Function Block

A function block contains all the requisite functions so that final processing of the measured value is done in this block. It is then ready for transmission to the control system. The blocks are designed to be independent of the sensor and the fieldbus. There are several function blocks available:

- **Analog input block:** It delivers the measured value from the sensor to the control system, i.e., Class 1 master after processing, simulation and proper scaling.
- **Analog output block:** It provides the device with a value as specified by the Class 1 master.
- **Discrete input block:** It provides a digital value from the device to the Class 1 master.
- **Discrete output block:** It provides the device with a value as determined by the Class 1 master.
- **Totalizer block:** It is used when a process variable is to be integrated, i.e., to be summed over a period of time. An example is a flow totalizer, the output value being collected by the Class 1 master.

8.9.3.4 Device Management Block

Today's devices contain a lot of information and can execute a number of functions hitherto done by PLCs and controllers. Such tasks can be executed correctly with the help of certain software tools needed for commissioning, maintenance, engineering, and parameterization of the devices.

PROFIBUS has developed the necessary software tools for device descriptions, ultimately leading to device management blocks. A device management block provides the information about the following:

- Which blocks are present in a device
- Number of blocks a device has and
- Where the starting address of a device is located

8.9.3.4.1 GSD

General station description (GSD) is a general and device-specific specification for communication. It is supplied by the manufacturer of PROFIBUS system. With the help of a keyword, the configuration tool can read the device identification, data type, and the permitted limit values from the GSD.

A GSD file can identify a PROFIBUS DP device (master/slave). It contains various information like vendor ID, baud rate supported, options/features supported, I/O signals and the length and timing of I/O. It permits plug and play, interoperability between devices from different manufacturers. The GSD file of each device is compiled into a master parameter record.

8.9.3.4.1.1 Specifications
There are three GSD specifications: general specification, master specification, and slave specification.

The general specifications contain device and vendor names, hardware and software versions, transmission rate, signal assignment on the bus connector, time interval to monitor times, etc. The master specifications contain master-related information like the maximum number of connectable slaves, upload/download options, etc. The slave specifications contain slave-related information like the number and type of I/O channels, diagnosis text specifications, and available modules in case of modular devices.

8.9.3.4.2 EDD

Electronic device description (EDD) is a very powerful software tool based on electronic device description language (EDDL). It can describe the application-related parameters and functions of a field device like configuration parameters, range of values, measurement units, default values, etc.

Using EDDL, a device manufacturer can create the relevant EDD files. These files provide the information to the engineering tool and the control system. Application areas of EDD include commissioning, runtime, engineering, asset management, documentation, e-commerce, etc.

8.9.3.4.3 FDT/DTM

Field device tool (FDT) is a manufacturer independent, open interface specification which helps in integrating field devices with the operator program using device type manager (DTM). FDT/DTM technology is subject to international standardization (IEC 62453).

Configuration and parameterization of field devices using existing languages have their own limitations like:

Intelligent field devices have their own complex, non-standard diagnostics which cannot be properly utilized by the existing software.

Both preventive maintenance and maintenance techniques are not properly included in the existing software.

These non-standardized tasks must be included in some *auxiliary tool* that would enable device manufacturers to integrate intelligent field device characteristics in the control system and at the same time allow the users with an expanded view of the field device characteristics.

FDT provides a universal interface that includes necessary software to address all the engineering and other automation requirements of the field devices. The specific functions of a field device like parameterization, configuration, diagnosis and maintenance, along with user interface, are mapped in a software component called device type manager. A DTM is a device operator program which helps in implementing either device functionality or communication capabilities. The manufacturer programs the DTM in a device-specific way and contains a separate user interface for each device.

8.9.3.4.4   ID

Every master and slave in a PROFIBUS system must have their individual IDs for proper system operation. On the basis of this ID, a master can identify the connected devices without having an extensive overhead. A master compares the ID number of a device with the ID number provided in the configuration data. Data transfer does not take place until and unless the correct device type having the correct address is connected to the bus. This will eliminate any possibility of configuration error.

8.10   NETWORK CONFIGURATION

It is set up with the help of GSD and configuration program provided by the manufacturer. A GSD file provides the following: vendor ID, features/options, baud rate supported, the length, and timing of I/O data. One major job of a GSD file is to support interoperability amongst different devices belonging to different manufacturers.

Each and every device belonging to the network has a GSD file of its own. These individual GSD files are loaded into the master parameter record. The master has an address allocation list of all the devices connected to the network.

During start-up operation, the master parameter record is used to set up communication with each assigned slave. When a new device is added to the network, it must be configured by the master so that it is inducted into the master parameter record. A system reset is performed before using the network. The master will try to establish contact with all the slaves connected to the network, before initiating data exchange. The process begins with the lowest address of the slave and ending with the highest one.

8.11   BUS MONITOR

Bus monitor, also called a protocol analyzer, is a software tool for monitoring and troubleshooting network activity. With its help, timing and packet content verification can be ascertained. A bus monitor is a PC which has a special PROFIBUS interface card and data capturing software.

Each captured message is time stamped with very high time resolution. The bus monitor is very helpful in indicating and diagnosing a problem occurring with an individual device.

8.12   TIME STAMP

Occurrence of certain important events and actions, like diagnosis and fault location, must precisely be known. Thus, such events must very precisely be *time stamped*, enabling precise time assignment for such events.

PROFIBUS has a time stamp profile for this purpose. This needs a clock in the slave devices—realized by a master clock in the system. An event can be given a precise time stamp and can be read accordingly. Messages are graded as per their priority with alarms falling under high and low priority. Figure 8.12 shows the technique of time stamping of alarms and events. The master reads acyclically such time stamped alarms and events from the buffer of the device.

8.13   REDUNDANCY

Redundancy ensures increased system availability. Redundancy can be applied at different stages like: master redundancy, media redundancy, segment coupler redundancy, ring redundancy, and slave redundancy.

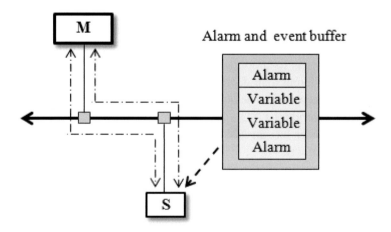

FIGURE 8.12   Time stamp and alarm messages. (Courtesy: PROFIBUS, Technology and Application: System Description, PROFIBUS International Support Center, Karlsruhe, Germany, Copyright by PNO 10/02, 2002.)

Master redundancy ensures the availability of a second master in case the primary one fails, i.e., the availability of the controller is ensured. Media redundancy ensures that the cabling is designed with redundancy. Segment coupler redundancy means that if a DP-PA gateway fails, the other will take over its function. Neither the master nor the slaves remain unaware of such a switchover. Ring redundancy ensures media redundancy on the PA side. Slave redundancy ensures installation of field devices with redundant communication.

The devices must have the following characteristics for slave redundancy to be ensured:

Slave devices must contain two independent Profibus interfaces. One is called the *primary* and the other is the *backup*.

Two independent protocol stacks must be there in the devices with a special redundancy expansion.

Within such a device, a *redundancy communication* (*RedCom*) must run between the two protocol stacks. Under normal conditions, communications take place over the primary slave, which also sends the diagnostic data of the slave. When the primary slave fails, either the secondary slave takes over seamlessly or is requested by the master to do so.

Slave redundancy can be realized on a single/two PROFIBUS line(s). The latter one ensures line redundancy. Slave redundancy provides high data availability, short reversing time, and no data loss and fault tolerant system.

8.14   PROFIsafe

There are several important issues that need addressing in serial bus communication. Some of these are: loss or repetition of data, delay in some data packet reaching the destination with delay, corrupt messages, incorrect sequences of data reception etc. PROFIBUS has designed PROFIsafe to counter these error possibilities.

PROFIsafe is a single channel software solution implemented in the devices above layer 7 of OSI. Figure 8.13 shows the fail safe mode with PROFIsafe.

PROFIsafe defines how fail safe devices like emergency stop pushbuttons can communicate over PROFIBUS with the help of failsafe controllers. PROFIsafe increases the transmission safety of the PROFIBUS protocol by incorporating the following features:

- Numbering of consecutive safety telegrams
- An identifier between sender and receiver in the form of a password
- Additional check on data in the form of CRC
- Timeout of incoming message frames and their acknowledgement

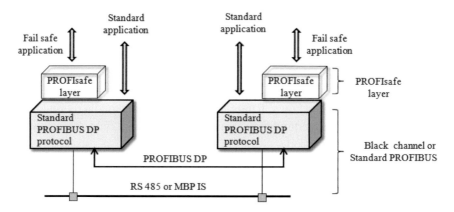

FIGURE 8.13 The fail safe mode as used with PROFIsafe. (Courtesy: PROFIBUS, Technology and Application: System Description, PROFIBUS International Support Center, Karlsruhe, Germany, Copyright by PNO 10/02, 2002.)

8.15   PROFIDRIVE

PROFIdrive is used in the area of electric drives. Its application ranges from simple frequency converters to highly sophisticated servo controls. The protocol defines six different application classes, depending on the complexity of the application.

Standard drive (class 1) is controlled by a set point value whereby the rotational speed is controlled by a drive controller. *Standard drive with technological functions* (class 2) involves dividing the main automation process into several smaller ones and then shifting some of the automation functions from the main controller to the drive controller. An example of class 2 type is the slave-to-slave communication. With *positioning drive* (class 3), an additional position controller is incorporated in the drive. An example is the twisting on and off of bottle tops. The positioning is precisely done by the drive controller. In *central motion control* (classes 4 and 5), multiple drives coordinate the motion sequence and the motion is controlled by a CNC machine. Other applications include motion control of linear motors. *Distributed automation* (class 6) has applications in slave to slave communication with the help of clocked processes and electronic shafts.

8.16   PROFIBUS INTERNATIONAL

PROFIBUS User Organization e.V. (PNO) was established in 1989 to promote PROFIBUS technology in manufacturing and process automation with a view to developing and maintain market dominance of the protocol. PROFIBUS International (PI) was established in 1995. PI has the following tasks to perform:

- Development and maintenance of PROFIBUS technology
- Enhancing acceptability of PROFIBUS technology worldwide
- Extending technical support for PROFIBUS technology through competence centers
- Quality assurance through device certification
- Enhancing members' interests through standardization committees and associations

8.17 FOUNDATION FIELDBUS AND PROFIBUS—A COMPARISON

A comparison between Foundation Fieldbus and PROFIBUS is shown in Table 8.4.

TABLE 8.4

Comparison between Foundation Fieldbus and PROFIBUS

Foundation Fieldbus	PROFIBUS
It is designed to address the needs of process automation.	It is designed to address the needs of discrete manufacturing and building automation.
Foundation Fieldbus H1 uses the physical layer as defined in IEC 61158-2.	PROFIBUS PA uses the physical layer as defined in IEC 61158-2.
It supports the requirements of intrinsic safety, including the new FISCO model.	It supports the requirements of intrinsic safety, including the new FISCO model.
A peer-to-peer protocol. Devices can communicate with each other without a host and they can initiate communications without a specific host command.	PROFIBUS PA is a master-slave protocol. A field device is a slave and it can only respond to a command from the master.
Control can be in the field device, or the host or partially in both.	With PROFIBUS PA, control resides only in the host.
When control is in the field devices, the host can be disconnected without halting the loop.	Host must be present for control to function.
It addresses interoperability by a combination of device descriptions and function blocks. A single host application can configure and access all device information and functionality.	A PROFIBUS PA host uses a standard profile for basic functionality. For additional vendor specific functionality, the host must have the corresponding software.

(Continued)

TABLE 8.4 *(Continued)*

Foundation Fieldbus	PROFIBUS
It uses device description technology to make all information available to all devices, host systems, and applications.	The device configuration and management host use device descriptions to configure and interact with the device. The control host uses profiles to access device information.
The device address can be manually/automatically assigned. It uses a special message to detect and identify a new device. A device can be added, deleted when the segment is in operation.	Device addresses are set by setting DIP switches or user entered software addresses. To add a device, the segment must be shut, and the device address and configuration parameters are configured in the host. The segment is then restarted.
It supports device and function block tags in the field devices. Thus, a device can be located by simply asking for it by its tag.	It supports tag in the host. Tag data base is manually entered in the host.
It provides a distributed real-time clock on the bus.	It does not provide a real time clock on the bus.
It is appropriate for real-time control on the bus—with or without a host.	It is appropriate for host based control only.

9 Modbus and Modbus Plus

9.1 INTRODUCTION

Modbus is a serial communication protocol developed by AEG-Modicon. It was initially designed to operate with programmable logic controllers (PLCs). It is an application layer messaging protocol, operating at layer 7 of open systems interconnection (OSI) protocol and provides client–server communication between devices connected on different types of networks. The Modbus protocol layers are shown in Figure 9.1, along with OSI protocol layers. It defines a method of accessing and controlling a device by another irrespective of the type of physical network involved.

For Modbus no interface is required, as is the case with many other buses. A user has the option of choosing between RS-422, RS-485 or 20 mA current loops. Compared to other buses, Modbus is relatively slower but has the decided advantage of very wide acceptance amongst control instrument manufacturers and users.

It is the industry's serial de facto standard since 1979. There is no formal way of certifying that a product is Modbus compatible. Rather, it is the responsibility of the manufacturers to confirm that their products are compatible with Modbus devices. The protocol describes the manner by

OSI layers		Modbus layers
7	Application	Modbus application protocol (client-server)
6	Presentation	Not used
5	Session	
4	Transport	
3	Network	
2	Data link	Modbus serial line protocol (master-slave)
1	Physical	RS 422, RS 488

FIGURE 9.1 The Modbus protocol layers.

which a device accesses another, how information is received, and how queries are responded to. In case of error, the protocol provides a mechanism to send the corresponding command to the user. Communication may take place on a Modbus network or on other networks (like Ethernet) by embedding the Modbus protocol as data packets in the protocol of the other networks.

The Modbus serial communication protocol is based on master-slave principle with the master initiating a transaction. The protocol provides for one master and up to 247 slaves.

Some characteristics of Modbus are fixed and some others are selectable by the users. The fixed characteristics are frame format, frame sequence, handling of communication errors and exception conditions and the functions performed. The selectable characteristics are transmission medium and transmission characteristics. The user characteristics, once set, cannot be changed when the system is in operation.

9.2 COMMUNICATION STACK

Modbus is implemented using the following:

- Transmission Control Protocol (TCP)/Internet Protocol (IP) over Ethernet.
- Asynchronous serial transmission over different media like: EIA/TIA-232E, EIA-422, EIA/TIA-485-A; fiber, radio etc.
- Modbus Plus—a high speed token passing method.

Figure 9.2 shows the implementation of Modbus communication stack using TCP/IP, master-slave, and Modbus Plus physical layer.

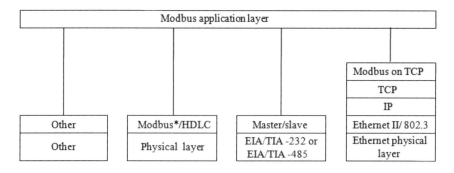

FIGURE 9.2 Modbus communication stack. (Courtesy: MODBUS, Application Protocol Specification, Vol. 1.1b, December 28, 2006. www.Modbus.IDA.org)

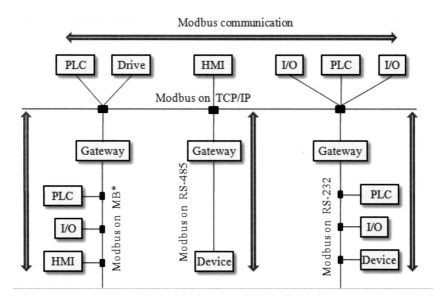

FIGURE 9.3 Modbus network architecture scheme. (Courtesy: MODBUS, Application Protocol Specification, Vol. 1.1b, December 28, 2006. www.Modbus. IDA.org)

9.3   NETWORK ARCHITECTURE

Figure 9.3 shows a scheme of Modbus network architecture. The different devices like PLC, human machine interface (HMI), inputs/outputs (I/Os), etc. can be connected to the Modbus TCP/IP via individual gateways. The different Modbus protocols, viz., Modbus on MB+, Modbus on RS-232, Modbus on RS-485 initiate remote communication by using TCP/IP.

9.4   COMMUNICATION TRANSACTIONS

Modbus serial communication uses the master-slave protocol. The master initiates the query and the slave responds by either providing the requisite data to the master or by taking the appropriate action as was requested for. The slaves respond by the following:

- by taking appropriate action
- by providing requisite data/information to the master
- by informing the master that the requisite action could not be carried out.

An error message, termed *exception response*, is sent to the master when the slave is unable to carry out the required actions as requested by the master. The exception response to the master contains the following:

- the address of the responding slave
- the action that the slave was requested to carry out and
- an indication of why the action could not be carried out.

A slave ignores a message if it contains some error. In such cases, the master resends the query to the slave, since it failed to receive a response from the slave.

9.4.1 MASTER-SLAVE AND BROADCAST COMMUNICATION

A master can individually address the slaves one by one (*unicast* mode) or addresses all the slaves at the same time (*broadcast* mode). The slaves respond to unicast message but do not respond to multicast messages. The master-slave communication and the broadcast communication modes are shown in Figures 9.4 and 9.5, respectively.

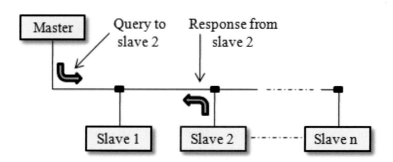

FIGURE 9.4   The master-slave communication model.

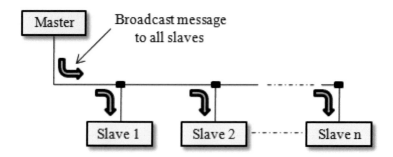

FIGURE 9.5   The broadcast communication mode.

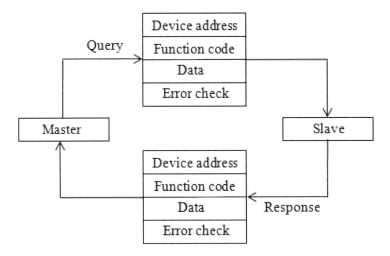

FIGURE 9.6   The query-response cycle.

9.4.2   QUERY-RESPONSE CYCLE

Query is initiated by the master and the slave responds to it. The query-response cycle is the basis of all communication transactions on a Modbus network. The query from the master contains four parts: device address, function code, data/message, and error check code. The response structure from the slave is identical to that of the master. Figure 9.6 shows query-response cycle.

9.4.2.1   Address Field

The address field sent by the master in its query contains the address of the slave for which the message is intended. Its value is in the range of 1 to 247 although practical limitations place a much lower value. When the slave sends its response, it places its own address in its address field so that the master can know that the correct slave is responding. Address **0** is earmarked for broadcasting. All slaves read them but do not provide any response to such query from the master.

9.4.2.2   Function Field

Function codes are in the range of 1 to 255, although not all the function codes are supported by all the devices. When a function code reaches a slave from a master in its query, the slave comes to know the actions that it will have to take. Examples of actions to be taken by the slave may include: read the input status, read register content, change a status within the slave, operating a relay coil, etc.

When the slave sends its response to the master, it repeats the function received. It indicates the slave has understood the query from the master and acted accordingly. If the instruction could not be carried out by the slave, it generates an *exception response* and the slave uses the function code and data field to inform the master the reasons for such exception.

In case the exception response is generated, the slave returns the original function code to the master, but with the MSB set to 1. The data field of the response message from the slave, in such cases, indicates to the master the nature of error that has occurred. Thus, the master can take appropriate actions based on this. The actions taken by the master may be either to repeat the original message or to try and diagnose the problem or to set an alarm etc.

9.4.2.3 Data Field

The data field received by a slave in the query from the master may typically include a register value, a register address or a register range. Some functions do not require the data field and thus the same is not included in the query from the master.

If no error has occurred, the data field of the response is used by the slave to pass data back to the master. When an error occurs, the data field from the slave passes on more information informing the master about the nature of error detected.

Modbus itself does not encode data and consequently a number of encoding schemes can be employed. This choice is left with the users of Modbus.

9.4.2.4 Error Check Field

The error check field allows the master to confirm the integrity of the message received from the slave. The error check method employed depends on the transmission mode selected. It may be cyclic redundancy check (CRC) for remote terminal unit (RTU) mode or LRC for ASCII mode of transmission.

On receiving the full message, the receiving device calculates the error check value and compares it with the error check value in the received message. If the two agree, no error has occurred and actions taken accordingly. The received message is rejected if the two values differ.

9.5   PROTOCOL DESCRIPTION: PDU AND ADU

Modbus can be implemented on different types of buses and networks, but the portion protocol data unit (PDU) is an integral part in each of them. It consists of two fields: function code and data.

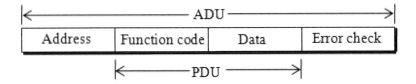

FIGURE 9.7   A general Modbus frame.

The two most common types of Modbus implementations are Ethernet (TCP/IP) and serial which may be RS 232, RS 422, or RS 485. The most common Modbus serial protocol is Modbus-RTU. Irrespective of the type of protocol used for a particular application, application specific addressing and error checking are attached to the PDU to give rise to application data unit (ADU). It is shown in Figure 9.7. ADU represents the Modbus frame. Since error checking is handled by TCP protocol, it is omitted in the ADU of a Modbus TCP.

9.6   TRANSMISSION MODES

Two types of transmitting modes are possible for transmitting serial data over a Modbus network: RTU and ASCII. They differ in a number of ways: the manner in which information is packed in the message field, the way the bit contents of the message are interpreted, the way the message is decoded, and the speed of operation at a given baud rate.

The different modes cannot be used together and the user has the option of selecting a particular mode for a certain application. The RTU mode is faster and more robust than the ASCII mode. Thus, it finds more applications than the ASCII form for message transmission.

The RTU transmission mode is sometimes referred to as Modbus-B while ASCII transmission mode as Modbus-A. The typical message length in ASCII mode is roughly twice the length of the equivalent RTU message. While in ASCII transmission mode, each byte in the message is transmitted as two ASCII characters, the same is sent as one 8-bit binary number containing two hexadecimal digits in RTU transmission mode.

Modbus packets can also be transmitted over local area networks and wide area networks by encapsulating the Modbus data in a TCP/IP packet.

9.7   MESSAGE FRAMING

For transmission of messages, a frame is constructed before its eventual transmission. A frame consists of start and end character(s), address of the device in the unicast mode or device addresses in the broadcast mode, function code, data and error check code.

Start	Address	Function code	Data	LRC	END
1 character	2 characters	2 characters	n characters	2 characters	2 characters

FIGURE 9.8   The ASCII frame format.

9.7.1   ASCII FRAMING

The Modbus ASCII frame consists of six fields as shown in Figure 9.8. The frame begins with a *start* which is a colon (:) character and it is ASCII character 3Ah, h signifying Hex. The frame ends with an *end*, which is a *carriage return-line feed* and representing two ASCII character pair 0D h and 0A h. The other remaining four fields in between have characters 0–9 and A–F in Hex.

ASCII mode allows an interval of up to 1 second between two successive transmissions without generating any error. All devices connected to the Modbus network continue to monitor the colon character which would start the start of an ASCII character. If a particular device finds the address matching with its own one, then it would start decoding the function code and the rest fields and takes actions accordingly.

9.7.2   RTU FRAMING

In this mode, the message frame starts with a silent time gap of at least 3.5 times character length. The message ends with the same time gap of 3.5 times character length. After the start field, the address field is monitored by the receiving devices to know of whether the message is meant for that device. The RTU frame format is shown in Figure 9.9. The entire message frame must be transmitted in one continuous stream or else an error will be generated.

9.8   MODBUS TCP/IP

The open Modbus TCP/IP specification was introduced in 1999. There are several advantages of using Modbus TCP/IP protocol like simplicity, use of standard Ethernet, openness, etc. Transfer rates in excess of 1 kB/sec can easily be achieved on a single station.

Start 3.5 character times	Address 8 bits	Function code 8 bits	Data N x 8 bits	CRC 16 bits	END 3.5 character times

FIGURE 9.9   The RTU frame format.

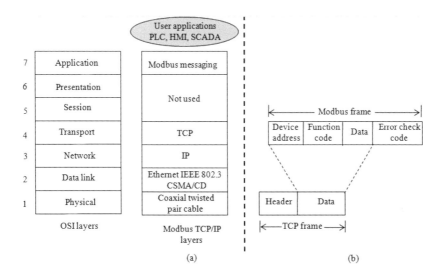

FIGURE 9.10 (a) Modbus TCP/IP layers (b) Modbus TCP/IP frame.

Modbus TCP/IP is an Internet protocol. It is simply a Modbus protocol with a TCP wrapper. Thus, Modbus devices can communicate over Modbus TCP/IP. A gateway device is simply needed to convert from the physical layer (RS-232, RS-485, or others) to Ethernet and to convert Modbus protocol to Modbus TCP/IP. Figure 9.10 (a,b) shows Modbus TCP/IP protocol layers along with OSI layers and also Modbus frame wrapped into the Modbus TCP/IP frame, respectively.

The normal Modbus frame contains the device address as either two ASCII codes (in ASCII mode) or an 8 bit hex byte (in RTU mode). The same is replaced by a combination of IP address and a unit identifier which identifies the device on the network. Also the 16-bit checksum in a Modbus frame is replaced by TCP's 32 bit CRC in the Modbus TCP/IP frame.

The master-slave architecture of Modbus protocol is modified to client-server architecture in Modbus TCP/IP. Since TCP is a connection oriented protocol, for every query in Modbus TCP/IP, there would be a response.

9.9 INTRODUCTION TO MODBUS PLUS

Apart from the standard Modbus protocol, there are two others: Modbus Plus and Modbus II, the latter one is less used because of additional cabling requirements. Modbus Plus is not an open standard protocol like the standard Modbus.

The *single master* limitation of Modbus protocol has led to the development of Modbus Plus which can share information and control strategy across various Modbus networks. Thus, Modbus Plus networks have individual Modbus communication networks. Modbus Plus was one of the early token passing protocols. It is a LAN in which the devices situated at geographically different positions can share information for measurement, control and monitoring purposes.

Modbus Plus caters to individual Modbus networks by bridging them together. The protocol allows a maximum of 64 devices on an individual protocol segment with each device assigned a network address in the range of 01–64. Messages can be routed from a device belonging to one Modbus segment to another by both the network address of the device and internetwork address. The routing of a message may be five layers deep.

9.10   MESSAGE FRAME

The Modbus Plus message frame is shown in Figure 9.11.

The frame begins with a preamble, followed by opening flag, broadcast address, data—whose length is variable in nature, an error check field and finally the closing flag. The variable data field itself has five fields: destination address, source address, MAC function, byte count, and LLC field.

Preamble	Opening flag	Broadcast address	MAC/LLC data	Error check field	Closing flag
1 byte	1 byte	1 byte	Variables	2 bytes	1 bytes

Destination address	Source address	MAC function	Byte count	LLC field (including Modbus command)
1 byte	1 byte	1 byte	2 bytes	Variables

Master output path	Router counter	Transaction sequence no.	Routing path	Modbus frame without CRC/LRC
1 byte	1 byte	1 byte	5 bytes	Variables

FIGURE 9.11   The Modbus Plus message frame. (Courtesy: S. Mackay et al., Practical Industrial Data Networks, Design, Installation and Troubleshooting, Newnes An Imprint of Elsevier, UK, p. 260, 2004)

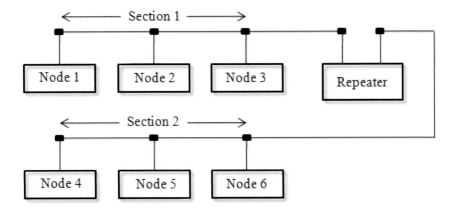

FIGURE 9.12   Two sections of a network connected by a repeater.

9.11   NETWORKING MODBUS PLUS

A single Modbus Plus network can support a maximum of 64 nodes. On a single section, up to 32 nodes (devices) can be connected with a maximum cable length of 450 m. The minimum cable length between any two successive pair of nodes is 3 m. Repeaters can be used to increase the node count to 64 in which case the maximum cable length becomes 1800 m. For longer cable distance requirements, fiber cables can be used. Figure 9.12 shows a single network consisting of two sections joined by a repeater. Thus, while a repeater is used to extend the length of a *single* network, a bridge is used to join *multiple* networks. A bridge cannot be used for deterministic timing of I/O processes.

A device on a network is assigned an address by the user of the network. No two device addresses can be same. The devices in a network act as peer members of a logical ring and a particular device identifies itself with the network on receiving the token frame. Each network maintains its own token rotation sequence. The token of a network is never passed on to another network. When a device or node holds a token, it can initiate message transmission with other nodes in the network.

When passing the token, a node writes into the global database which is then broadcast to all the other nodes in the same network. The global data represents a field in the token frame. Nodes in the network monitor the global data and extract the same for its own use like updating of alarms, set points, etc. Each network in a multiple network system has its own global database, as the token is not passed from one network to any other.

A typical global plus network with dual cabling is shown in Figure 9.13. The two cables are termed as cable A and cable B. Each cable length will have a maximum length of 450 m. The maximum difference in length

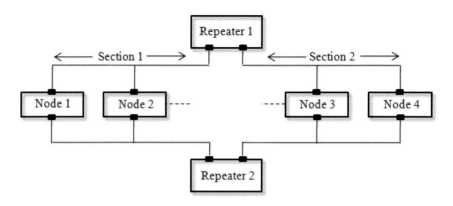

FIGURE 9.13   A Modbus Plus network with dual cabling.

between cable A and cable B cannot be more than 150 m, measured between any pairs of nodes on the cable section.

The token passing mechanism works like this: the token rotation begins with the network's lowest addressed active node, increases progressively until the highest active node address is reached. The token then begins at the lowest addressed active node again.

Addition and deletion of nodes to a network are transparent to the user. When a node leaves a network, a new token passing sequence will be established in 100 ms. On the other hand, if a new node joins a network, it will be added to the newly formed address sequence in 5 seconds.

10 CAN Bus

10.1 INTRODUCTION

CAN stands for controller area network and was developed by BOSCH of Germany in 1986 to take care of the growing demand of electronic control systems in automobile industries. CAN is a serial communication bus protocol standardized by International Standards Organization (ISO). It does not use master-slave or token passing method to access the bus. Instead it uses a unique bus access control method, called *non-destructive bit wise arbitration* to access the bus. It is a very simple, highly reliable, and prioritized communication protocol amongst sensors, actuators, and intelligent devices. A producer/consumer technique is applied to access the physical medium based on carrier sense multiple access with collision detection (CSMA-CD). It is a deterministic method to resolve collision/conflict on the bus by taking recourse to bus contention. The system thus uses the full bandwidth of the medium.

CAN refers to a network of independent controllers. It supports distributed real-time control with a very high level of security.

The different controls in an automotive vehicle are of different data types requiring multi bus lines to be sent to the controller. This resulted in many wires leading to various problems. CAN was developed to effectively address the above and became a standard for vehicle networking. It has since been applied to numerous fields for control purposes.

10.2 FEATURES

The units connected to the bus can send messages when the bus is free, i.e., the system is multimaster type. When more than one unit starts sending messages at the same time, their priority is resolved by a message identifier residing in the data frame. Thus, a particular unit wins the bus contention and message is sent by that unit. The other units which lost out, can send their messages when the bus goes into idle state. The units connected to the bus do not have any address and any unit can be added/deleted at any time without any change in software or hardware. The speeds of the connecting units in the network are unique as already set depending on the network size.

Data from outside the network can be accommodated by sending a *remote frame* to the units. CAN protocol has error detection, error notification, error recovery, and error confinement capabilities.

Number of units which can be connected to the system has no logical limit, although it is dependent on latency and load on the bus. When more units are added to the system, speed decreases and vice versa.

10.3 TYPES

CAN has several ISO standards like ISO 11898 and ISO 11519-2. These two standards do not differ in data link layer but have differences in physical layer.

ISO 11898 refers to high speed CAN communication. Later on, it is divided into ISO 11898-1 and ISO 11898-2. The former standing for the standard for data link layer while the latter for physical layer. The standard ISO 11519-2 refers to low speed communication with speeds up to 125 kbps.

10.3.1 SPEED VS. BUS LENGTH

Communication speed decreases with bus length. As the bus length increases, the communication speed decreases and shown in Figure 10.1.

10.4 CAN FRAMES

There are five types of frames for communication in CAN. These are: data frame, remote frame, error frame, overload frame, and interframe space. The first two frames are set by the user while the rest are set in hardware portion of CAN.

The data and remote frames come in two formats: standard and extended. The standard has an 11-bit ID while the extended version has a 29-bit ID.

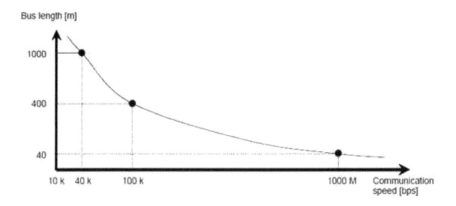

FIGURE 10.1 Communication speed vs. bus length. (Courtesy: Renesas Electronics America Inc. Introduction to CAN, REJ05B0804-0100/Rev. 1.00. USA, 2006.)

If a device wants to know the data associated with an identifier which it does not know, then the device sends a frame, called a remote frame, which has the desired identifier in its arbitration field with the RTR in set condition and the data field empty.

10.5   CAN DATA FRAME

The most important frame amongst the five frames in CAN is the data frame and is shown in Figure 10.2.

The data transmit unit uses the data frame to send a message to the receive unit. The data frame consists of seven fields: start of frame, arbitration field, control field, data field, CRC field, ACK field, and end of frame. The number of bits in each field is shown in the figure. The data field is of variable length—it can be a maximum of 8 bytes. This is sufficient to convey the information requirements from most of the devices. CRC field in conjunction with ACK field lends integrity to data as it is sent over the CAN bus.

10.6   CAN ARBITRATION

The bus arbitration principle about the device which gains control of the bus is shown in Figure 10.3. The device which first outputs a message on the bus during a bus idle state gains control of the bus and sends information through it. In case more than one device wants to send data over the bus, a station identifier bit pattern for each device try to gain access of the bus.

Priority regarding access of the bus by a device is determined by the addressing assignments during configuration of the network and allows the highest priority device to access the same. The devices lower in priority lose the contention and get another chance once the bus becomes idle.

The data frame consists of a station identifier field of 11 bits and a single remote transmission request (RTR). The identifier field determines the priority of a device over others.

In the figure, three devices try to gain control of the bus at the same time. State 0 is *dominant* while state 1 is *recessive*. The **0** state dominates over the **1** state. The devices are connected to the network by an open collector stage. The bits of devices 1, 2, and 3 as configured during initialization are put on the bus and shown in the figure. Device 1 first loses bus arbitration due to both 2 and 3. After that, device 3 loses out to device 2. The pattern on the bus is thus corresponding to the identifier bit pattern of the device which wins the contention, which in this case is device 2. Devices 2 and 3 again get the next chance to transmit a message once the bus becomes idle. In conclusion it can be said that if a device while transmitting a recessive bit detects a dominant bit from another device, the former stops transmitting. In this way, the less

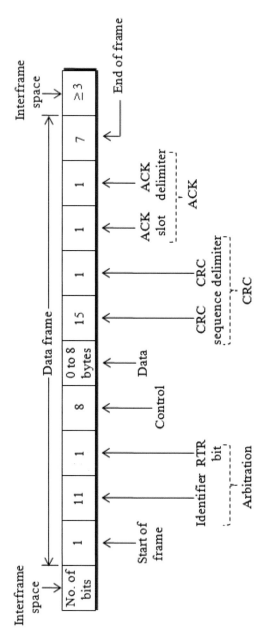

FIGURE 10.2 The CAN bus data frame. (Courtesy: Practical Data Communications for Instrumentation and Control, J. Park. et al., p. 266, Newnes, 2003)

RTR: Remote transmission request, CRC: Cyclic redundancy code, ACK: Acknowledge

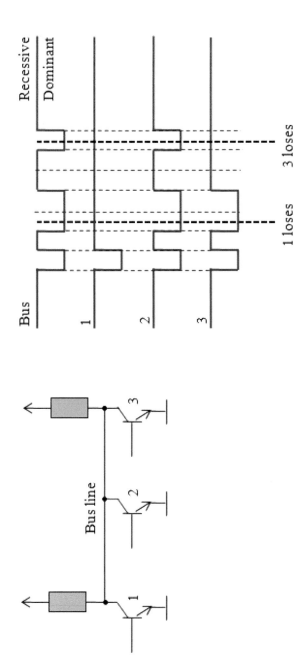

FIGURE 10.3 The bus arbitration principle in CAN bus. (Courtesy: Practical Data Communications for Instrumentation and Control, J. Park. et al., p. 266, Newnes, 2003)

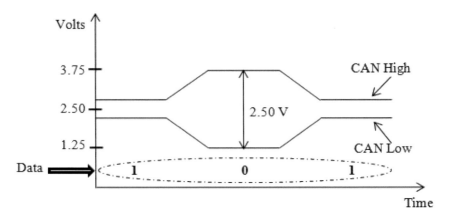

FIGURE 10.4   Voltage levels in CAN bus communication.

prioritized devices lose the contention and the highest prioritized device wins the same, thereby gaining the right of way over the bus.

10.6.1 CAN COMMUNICATION

When not communicating, the CAN bus is idle and both the wires are tied at 2.5 V. Two dedicated wires are used for communication. The wires are called CAN high and CAN low. When a message is being transmitted, CAN high assumes 3.75 V and CAN low voltage becomes 1.25 V, causing a voltage gap of 2.5 V between the two lines. This is shown in Figure 10.4. Since communication relies on voltage differential between the two lines, it is robust in networked communication of its insensitiveness to inductive spikes, electromagnetic noise, etc.

10.7 TYPES OF ERRORS AND ERROR STATES

There are five types of errors that may occur in CAN communication, with more than one error may occur simultaneously.

A device remains in one of three error states, which are: error active state, error passive state, and bus off state. In the error active state, a device can participate in communication on the bus in a normal manner. In case a device detects an error in the error active state, it sends an active error flag. A device tends to cause an error in the error passive state. In this state, the device does not communicate to other devices about the error in the receive mode. A passive error flag is transmitted when the device detects an error in the error passive state. In the bus off state, a device can't participate in communication. It is disabled from all transmit/receive operations.

11 DeviceNet

11.1 INTRODUCTION

Developed by Allen-Bradley in 1994, DeviceNet is a device or low-level industrial open network that communicates between the device controller and sensors or actuators like limit switches, valve manifolds, motor starters, variable frequency drives, remote I/Os, etc. It is included in EN 50325 and IEC 62026 standards. DeviceNet is a CAN-based layer 7 application protocol and is maintained by Open DeviceNet Vendor Association, Inc. (ODVA). The association issues specifications and ensures compliance with the stated specifications.

Devices from different manufacturers that comply with DeviceNet standards can be connected together in the network. The DeviceNet specification is defined in two volumes: Volume 1 and Volume 2. Volume 1 pertains to application layer which uses common industrial protocol (CIP), data link, and physical layers, which use CAN protocol. Volume 2 is concerned with the device profiles to obtain interoperability and interchangeability among the different products.

11.2 FEATURES

DeviceNet supports up to 64 nodes with a maximum device count of 2048. Network topology used is trunk or bus line with drop cables which connect to the devices. On either end of the trunk line, terminating resistances of value 121 Ω are placed. It supports use of repeaters, bridges, routers, and gateways.

DeviceNet is used when devices are mostly discrete in nature with an analog mix and where motor control and variable frequency drives are present. It has moderate transmission speed and it depends on cable run lengths.

DeviceNet supports master-slave, peer-to-peer, and multi-master modes to transfer information in the network, utilizing the total bandwidth in the process. Slaves can be owned by one master only.

DeviceNet is robust in nature offering some diagnostics at the same time. It has the capability to detect duplicate node address. It supports power and signal on the same cable. DeviceNet supports 8 Amps on the bus. Because of huge power handling, the system is not intrinsically safe. Devices can be added/withdrawn from the network under power.

DeviceNet network is classified as a device bus network. Its characteristics are byte-level communication, high speed, and a lot of diagnostic power by the devices in the network.

Every DeviceNet device has a configuration file in its electronic device data sheet (EDS file). It maintains important data about the device which must be registered on the network configuration software. A configuration tool is used to input the EDS files and configuring the devices. Devices which are modularly designed use one EDS file per component, i.e., module. This definitely is advantageous in configuring such devices.

11.3   THE OBJECT MODEL

A DeviceNet station or node is considered as a collection of objects. An object provides an abstract representation of a particular component in a product. Class is a group of objects that represents the same type of component while the attributes provide the characteristics of objects.

Every device has mandatory and optional objects. Optional objects give a device the category (also called profiles) to which it belongs like pneumatic valve, AC/DC drive, etc.

11.4   PROTOCOL LAYERS

Figure 11.1 shows the DeviceNet protocol along with OSI protocol. It is seen that layers 3 to 6 of OSI protocol are absent in this case. DeviceNet

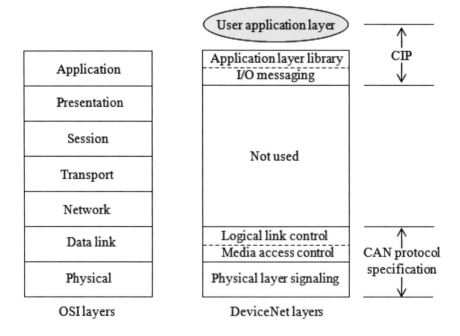

FIGURE 11.1   DeviceNet protocol layers.

uses CAN protocol to implement data link and physical layers while application layer conforms to CIP.

11.5 PHYSICAL LAYER

DevicNet uses CAN protocol for its physical layer implementation and it offers a great advantage because of easy availability of CAN chips. Each node in the network is assigned a unique address in the range 0–63. The same is set by using switches on the devices and a configuration tool existing in the system.

The devices can be powered from the bus line or else separately. Each device power should at least be 11 V or higher. For greater protection, optical isolation is taken help of.

11.5.1 DATA RATE

DeviceNet protocol offers variable data rate, trunk distance, and drop length. This relationship is shown in Table 11.1. The total length of trunk line is dependent on data rate, cable type, and the number of devices. Devices can be connected by means of screw terminals, sealed type screw tight connectors, etc.

11.6 DATA LINK LAYER

Data link layer comprises of logical link control (LLC) and media access control (MAC) and follows data link layer of CAN specification. As already mentioned, carrier sense multiple access with collision detection (CSMA-CD) is used to access the physical medium. Contention on the bus is resolved by a priority based deterministic procedure. This avoids collision on the bus and at the same time utilizes the total bandwidth of the transmitted medium.

TABLE 11.1

Relationship between Data Rate, Trunk Distance, and Drop Length

Data rate	Trunk distance (m)	Drop length Maximum (m)	Drop length Cumulative (m)
125 Kbaud	500	3	156
250 Kbaud	200		78
500 Kbaud	100		39

It should be mentioned here that although DeviceNet uses the CAN data frame, encoding of identifier and the data packets for DeviceNet is left to application layer developer. A producer/consumer model is used for communication on the bus, where a station places data on the bus (producer) and this is read off by another station on the network (consumer).

11.7 APPLICATION LAYER

DeviceNet categorizes CAN frames into four groups, from group 1 to group 4, with decreasing priority levels. This is achieved with the help of an 11-bit CAN identifier field. It includes a message ID and a MAC ID. The MAC ID can either be a source or a destination.

A predefined master/slave connection set defines the group via which data transfer is to take place. Such an action is done to considerably reduce software load to establish a connection. To this end, the master sends a service request to the designated slave(s). A subsequent request by the master establishes the connection which then takes the help of the predefined master/slave connection set.

CAN protocol, used by data link and physical layers, does not interpret the fields in the CAN message frame. In the case of DeviceNet, software developed and residing in the application layer distinguishes between two types of messages: cyclic I/O and explicit type. Cyclic I/O is of four types: bit strobe command/response message, poll command/response message, change of state message, and cyclic message. The type is dependent on the manner in which data is exchanged. These are now discussed:

- *Bit/strobe command* is used for exchange of small amount of data. Here, the message gives out 8 bytes, i.e., 64 bits of data, one for each slave. If a slave is absent or not included in the scan list, the corresponding bit does not carry any meaning. The response from the slaves can contain a maximum of 8 bytes of data.
- In the *poll command* type, the master sends a message to a single slave. The slave may also return data to the master.
- In the *change of state method* of communication, data exchange may be initiated by either master or slave. This takes place if the values monitored or controlled change within a set time limit. This time limit is set in the network configuration program.
- In the *cyclic method*, data exchange takes place at regular time intervals regardless of being altered or not. These intervals are adjustable on the network configuration program.
- *Explicit Message* type is a general purpose one. This is used in asynchronous jobs, e.g., to transfer the values of some attributes, parameterization, and configuration of equipment.

TABLE 11.2
Error States

Non existent	The device has shut down due to an internal error or some remote command
Unallocated	The device has successfully joined the network but is not currently owned by a DeviceNet master device. The LED network status indicator for an unallocated device flashes green
Timed out	Messages have failed to arrive at one or more connections with the master device. This is typically a recoverable error. The network status LED indicator on a device will flash green in this state
Faulted	The device has detected an internal error or received a duplicate MAC ID response message. This is not a recoverable error. The network status LED indicator on a device will typically be solid red in this case
Bus off	In the BUS off state, the device has detected significant network errors and has removed itself from network operation. This is typically a hardware failure in the device circuitry. The network status LED will typically be solid red in this case

11.8   POWER SUPPLY AND CABLES

A single four conductor cable in bus topology configuration can be used in DeviceNet. Two wires are used for supplying power to the devices and two for supporting communication. Both the pairs have a foil shield. The cable has an overall braiding. ODVA recommends the use of different types of cables as follows: ODVA Type I thin cable for drop wires, ODVA Type II thick cable for trunk wire while ODVA Type III is used when flexible drop cable is the requirement. 24 V DC power is provided on the power lines and supports 8 A on the thick cable and 3 A on the thin cable. The thick and thin cables are connected together by T junctions.

11.9   ERROR STATES

DeviceNet devices assume any one of the following error states, shown in Table 11.2.

12 AS-i

12.1   INTRODUCTION

Actuator sensor interface (AS-i) is a bit-oriented master-slave type open system fieldbus designed for use at the device level of process control systems. It is included in both EN 50295 and IEC 62026 standards. AS-i is designed to connect binary sensors, actuators (i.e., slaves) which require very small number of bits to convey device status. It cannot be efficiently used with intelligent controllers which require information beyond the limited capacity of such a protocol.

Design of AS-i is based on modular components. They act as bridges between network and binary sensors. Nodes can be added or taken out in *live* condition, i.e., while the system is running. AS-i data capacity ranges from 1 to 16 bits per device.

12.2   FEATURES

It is a cyclic polling, single master multi-slave type system where the slaves have specific addresses and sensors/actuators from different manufacturers can be connected together. Topology used can be ring, linear, star, or tree with a cable length of 100 m which can be extended to 300 m by using repeaters. The *alternating pulse modulation* technique used during transmission of information reduces the bandwidth and also the *end of line reflection* is very common in networks that use square wave pulse techniques. Synchronizing information is passed on to receiver giving AS-i protocol a very reliable one as far as data integrity is concerned. It is an open standard in which error detection and retransmission of incorrect data is possible. It eliminates the need to employ PLC input and output modules, hence cost is low per slave basis. Data and power for the devices are sent through two unshielded, non-twisted cables. Each slave typically consumes 200 mA with a maximum 8 A per bus. Cycle time is a maximum 5 ms for version 2.0 and 10 ms for 2.1 versions.

12.3   DIFFERENT VERSIONS

AS-i protocol is available in three versions: the original Version 2.04 (1994), Version 2.11 (1998), which is an enhanced version over the earlier one, and Version 3.0 (2005/2007), featuring some additional capabilities.

These versions are also known as AS-i 2.0 or AS-i 1 specification, AS-i 2.1 or AS-i 2 specification and AS-i 3.0 or AS-i 3 specification, respectively.

In version 2.04, a maximum of 31 slaves could be employed with each slave linking four digital inputs and four digital outputs resulting in 124 inputs and 124 outputs on a single network. A feature of this version is the automatic substitution of a network module. Update time is approximately 150 ms. This is calculated by multiplying the number of input/output (I/O) nodes with the deterministic update time for each node.

In version 2.11, number of slaves increased to 62 from 31, number of I/Os to 434, and the update time to 10 ms. The number of slave profiles increased to 225 from 15 and the peripheral error could be taken care of by adding a bit in the status record.

In version 3.0, some of the extra capabilities added are: full duplex bit serial data channel and configurable fast analog channel of 8, 12, or 16 bits.

12.4   TOPOLOGY

Communication in AS-i is controlled by a single master, with which the slaves are connected in various configurations, like star, line, branch, or tree. These are shown in Figure 12.1.

12.5   PROTOCOL LAYERS

Figure 12.2 shows the AS-i protocol along with the OSI protocol. Layers 3–7 of OSI are non-existent in this case. Layer 1 is termed as *transfer physics* while layer 2 is called *transmission control*. There is a layer, called

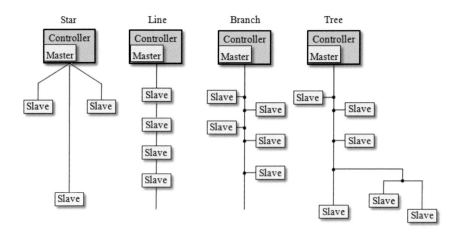

FIGURE 12.1   Physical network topologies.

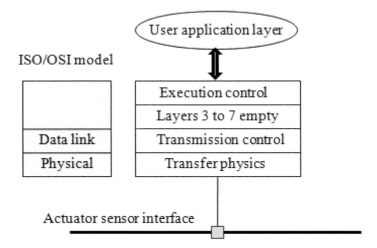

FIGURE 12.2 AS-i protocol. (Courtesy: B. G. Liptak. Instrument Engineers' Handbook, Process Software Digital Network, 3rd Edition, CRC Press, Boca Raton, FL, p. 621, 2002)

execution control, and placed above layer 7 of OSI, which is responsible for overall operation of AS-i.

12.6 PHYSICAL LAYER

The physical layer is responsible for physically establishing the connection between master and slaves. It uses a two wire untwisted and unshielded cable, which is available in two versions and is used for both communication and power supply to the devices, shown in Figure 12.3. The general cable version is used to connect devices by normal screw terminals while the special cable can connect stations directly by contacts that penetrate the cable isolation.

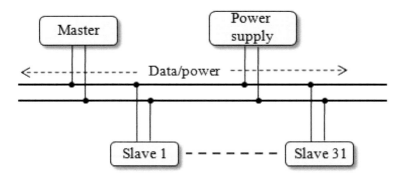

FIGURE 12.3 Master slave connection in AS-i.

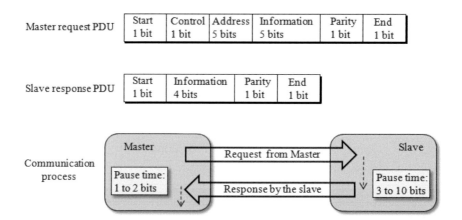

FIGURE 12.4 AS-i master and slave PDUs and communication sequences. (Courtesy: B. G. Liptak. Instrument Engineers' Handbook, Process Software Digital Network, 3rd Edition, CRC Press, Boca Raton, FL, p. 622, 2002)

12.7 DATA LINK LAYER

It consists of a master call-up and a slave response. It is a bit-oriented protocol with the master call-up, shown as master request PDU in Figure 12.4. It has six fields. The slave response is shown as slave response PDU. After master request, there is a pause of 3–10 bits, while it is 1–2 bits pause for the slave request PDU (to master). This communication sequence between request-response is shown in the lower part of the same figure. The pause between each transmission is used to ensure synchronization, error detection, and error correction.

Various combinations are possible in the 5-bit information field of the master PDU, the content of which tells the operation to be executed on the slaves. Some of them are Address_Assignment, Data_Exchange, and Write_Parameter. When the code corresponding to Address_Assignment is set, the master sets the address of the slave, which is possible only if the slave has a default address that does not allow for data exchange.

12.8 EXECUTION CONTROL

Execution control layer exists at the top of layer 7 of AS-i. It is necessitated for the management and overall proper operation of AS-i. The user application layer requests the execution control layer for different functionalities. The main jobs performed by execution control unit are: initialization, start-up and normal operation.

Initialization involves setting some parameters offline, like setting some parameter values for the master, testing the power supply to ensure whether it can adequately supply the load to the slaves.

In the start-up phase, the master detects and activates the slaves. There are two ways to achieve the above: the protected operation mode and the configuration mode. In the former, user prepares a list beforehand, called *list of projected slaves* and the master activates the slaves only from that list only. In the latter, all the slaves detected by the master are activated.

The jobs carried out by the normal operation phase include: cyclic data transfer between master and slave, acyclic management tasks and inclusion function. In cyclic data transfer, the master updates each slave with output data cyclically and acquires data from each slave in an identical fashion. This cyclic data transfer automatically takes place without user intervention. User requests are taken care of in management tasks. Inclusion function allows inclusion of some slaves to the cyclic data transfer. For this, the master first detects a new slave and then only the slave can be activated for cyclic data transfer to take place with this new slave.

12.9 MODULATION TECHNIQUE

Alternating Pulse Modulation (APM) technique is used for data transmission in case of AS-i, shown in Figure 12.5. It produces a baseband signal which is superimposed on DC power supply. As information field of AS-i is of limited in size, APM is used to ensure data integrity. The modulated signal is *sin squared*, which is similar to Manchester II coding. This coding scheme reduces the bandwidth required in transmission and end of

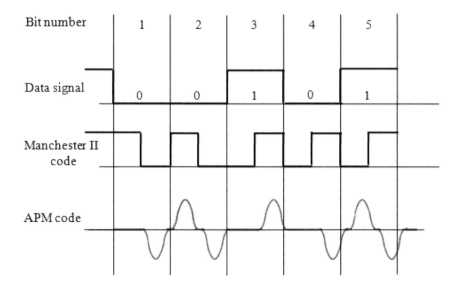

FIGURE 12.5 APM technique used in AS-i.

line reflections. As shown from the figure, each data bit has an associated pulse in the second half of bit period which helps in bit-level error checking. Also this scheme helps in synchronization at the receiving end. The start or first bit of APM is a negative pulse while the stop bit is a positive pulse. Even parity and a recommended frame length are included in APM technique. These inclusions offer more credence to data integrity in the received signal.

13 Seriplex

13.1 INTRODUCTION

Seriplex control bus, developed by Automated Process Control Inc., in 1987, is a control network specifically intended for simple industrial control applications. Seriplex is most favored for simple devices like push buttons, valves, contactors, limit switches, etc. A single cable connects all the devices to save considerably on cabling costs. The total cable run is about 1500 m while the same for other networks is about 300 m.

Seriplex control schemes employ intelligent modules which provide a link between inputs and outputs, much alike logic gates. This obviates the need to have a supervisory processor. For supervisory control systems, the network is to be connected to a host processor via interface adapters.

Various topologies are available for connecting the intelligent modules to the Seriplex network via cables which provides power for data and clock signals. Without address multiplexing, Seriplex can support 255 input bits and 255 output bits; sixteen 16-bit analog inputs and 16-bit analog outputs. With address multiplexing, the system can support 3840 input bits and 3840 output bits; two hundred forty 16-bit analog inputs and two hundred forty 16-bit analog outputs

Seriplex bus can be configured to operate in both master-slave and peer-to-peer versions.

13.2 FEATURES

A lot of features have gone into the development of Seriplex control bus. It is highly cost-effective for networking simple control devices. Seriplex protocol is very efficient for data length ranging from 1 to 64 bits while other protocols are efficient for data length in excess of 16 bits. Thus, for very simple control systems, Seriplex protocol is mostly used. The bus is inherently deterministic, implying that the time at which response is available is known. Seriplex employs a highly efficient protocol with less overhead. Hence, its response is very fast although its baud rate is less than other protocols.

Without multiplexing, up to 510 bits can be transmitted in frames continually with very short synchronization periods. The data bits located within the frame corresponds to the addresses assigned to the I/O devices.

The data and clock signals operate at 12 V with a 4.5 V respectively. This gives very good noise immunity. The network power cable is separate from I/O power cable. Additionally, two other wires are earmarked for data and clock signals. The four conductors are enclosed in a shield.

Efficiency of Seriplex systems can be as high as 98 percent. It means that for 98 percent time the system transports actual data bits. The setup tool available in Seriplex systems is meant for configuring master-slave mode and it takes around 15 seconds to configure them.

Seriplex control bus updates all I/O data on a regular basis, thus there is no need to have a collision resolution or message prioritization

13.3 PHYSICAL LAYER

Physical layer, belonging to layer 1 of open systems interconnection (OSI), connects Seriplex network with each individual device. It is a four conductor cable, with two thicker wires of AWG # 16 meant for power and common. The relatively two thinner wires of AWG # 18 carry data and clock signals. Maximum clock rate is 200 kHz.

First generation I/O devices are powered by 12 V DC while the second generation ones are powered by either 12 V or 24 V DC, the value employed depends on user requirements. Connections are made through Seriplex modules placed near field devices.

Cable capacitance plays a vital role in determining the maximum distance that data can be delivered. The lower the cable capacitance, the greater the distance. A cable capacitance of approximately 50 pF/m allows data rates up to 100 kHz for a maximum distance of 150 m while a cable capacitance of 65 pF/m would enable data transmissions at 100 kHz for a maximum distance of 100 m. Number of digital and analog I/Os depends on address multiplexing.

13.4 DATA LINK LAYER

Seriplex control bus can be operated in two modes: peer-to-peer mode and master-slave mode. In the former, no host controller is needed and data can be interchanged between devices, while for the latter, a host controller along with an interface card controls the activities on the bus following some preloaded software.

Peer-to-peer timing diagram is shown in Figure 13.1. This is called Mode 1. Each input and output device are assigned the same address. Data with an input address 7 appears as output with the same address 7. A separate clock is required in this case as no host is there to provide the same.

In the master-slave mode, also known as Mode 2, there are two clock pulses per address and the timing diagram is shown in Figure 13.2. During the first clock pulse for an address, input data is transferred to the interface

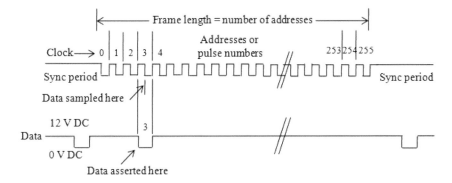

FIGURE 13.1 Seriplex peer-to-peer timing diagram. (Courtesy: Square D. SERIPLEX: Design, Installation and Troubleshooting Manual, Bulletin no. 30298-035-01 Raleigh, NC, 1999. www.guilleviniag.com/downloads/Products/Schneider/Seriplex.pdf.)

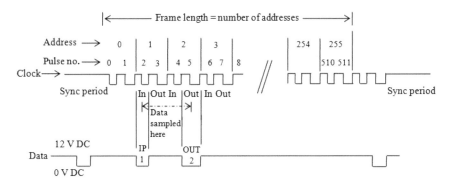

FIGURE 13.2 Seriplex master-slave timing diagram. (Courtesy: Square D. SERIPLEX: Design, Installation and Troubleshooting Manual, Bulletin no. 30298-035-01 Raleigh, NC, 1999. www.guilleviniag.com/downloads/Products/Schneider/Seriplex.pdf.)

card and from there it is retransferred to the host. During the second clock pulse for the same address, the host gives out output data to remote devices. This method thus facilitates separation of input and output data for the same address. In this mode, the host takes all the control decisions regarding data transfers on the bus.

13.5   DATA INTEGRITY

Seriplex systems offer a fair degree of data integrity in the form of *digital debounce* and *data echo* features. These two methods efficiently verify data for its integrity than either parity or CRC method of data verification for its authenticity. The second generation Seriplex ASIC2B provides the above two features.

Devices such as valves and contactors which receive data, check data stability for multiple readings, before acting upon in order to filter out spurious data.

Data echo mechanism can determine whether some device is actually connected or not and whether the ASIC is operating properly or not. For output data from output devices, data can be echoed and monitored. If data is properly echoed, it can be concluded that the device is operating faithfully and the ASIC is responding properly to data transmissions.

Multi-bit data such as obtained in case of analog signals, can be verified by employing complementary data retransmission (CDR) feature which is supported by Seriplex version 2.

14 Interbus-S

14.1 INTRODUCTION

Interbus-S is an open architecture, serial bus, mater-slave type fieldbus communication system that operates in an active ring topology. It consists of one master and up to 511 slave devices. Each slave has an input and an output connector. It was developed by Phoenix Contact in 1984 to interchange data between control systems like PCs, PLCs, and distributed I/O modules that communicate with sensors/actuators. A maximum of 4096 number of digital I/O points for a maximum distance of 400 m can be connected via this network. These points can be updated in 14 ms by employing a variable frame transfer protocol. It can be reduced by employing lesser number of I/Os. Data transfer rate on the bus is 500 kbps. The basic scheme of process control via Interbus is shown in Figure 14.1.

A bus cycle begins with the master passing a bit stream to the first slave. The first slave then transfers the same to the second one in the ring and at the same time data from the first slave is transferred to the master.

14.2 FEATURES

The protocol is efficient, deterministic, cyclic, and full duplex in nature. The network system is manufacturer independent and can adapt to future modifications and expansions. Because the devices are connected in a ring fashion to transfer data from each device to the next in the line, Interbus

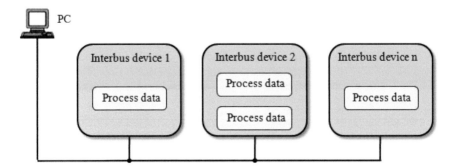

FIGURE 14.1 Basic scheme of process control via Interbus. (Courtesy: Dr.-Ing. Reinhard Langmann, INTERBUS Basics. Process Control Laboratory, University of Applied Sciences, FH Dusseldorf.)

works as a network based shift register. It can handle both single bit data (process data) as also data records for intelligent field devices (parameter data). Process data is transmitted in fixed and cyclic time cycles in real-time mode while large volume parameter data is transmitted in acyclic transmission. Data varying between 1 and 64 bytes per data direction is permitted in a single Interbus device.

14.3   OPERATION

Interbus is a master-slave type communication system. The master acts as a controller with all the devices connected in a ring fashion. Each slave, i.e., device has two distinct lines for data transmission: one for forward and the other for return data transmission.

Figure 14.2 shows the basic structure of an Interbus system. It consists of one main ring and sub-rings—also known as bus segments. Each sub-ring is connected by a bus coupler (or bus terminal module).

The local bus is connected to local bus devices, while the remote bus is connected to remote devices and bus couplers. It is controlled from the

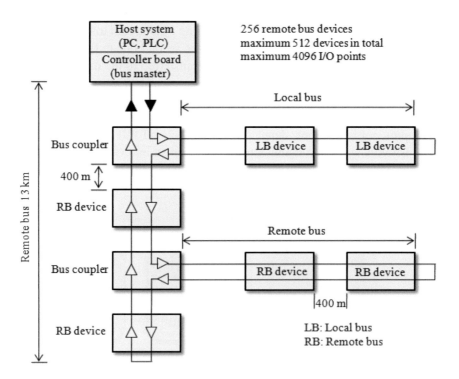

FIGURE 14.2   Structure of an Interbus system. (Courtesy: Dr.-Ing. Reinhard Langmann, INTERBUS Basics. Process Control Laboratory, University of Applied Sciences, FH Dusseldorf.)

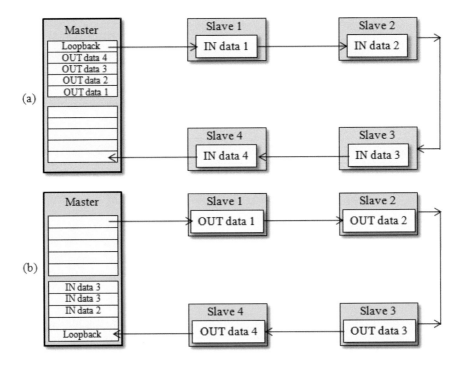

FIGURE 14.3   Principle of data transmission on Interbus: (a) Distribution of data before a data cycle. (b) Distribution of data after a data cycle. (Courtesy: Dr.-Ing. Reinhard Langmann, INTERBUS Basics. Process Control Laboratory, University of Applied Sciences, FH Dusseldorf.)

controller board. Each bus coupler is connected to a sub ring. The remote bus branch is connected to remote devices spread over a wide geographical area. It is suitable for complex networking for processes distributed over a wide area, separated by large distances.

The process of sending data from the master to the slaves (OUT Data) and from the slaves to the master (IN Data) is executed in full duplex transmission mode and is shown in Figure 14.3, in which (a) and (b) correspond to data positions before and after a data cycle. The master provides a data packet, contained in a summation frame, to the send shift register. The data registers in each device (slave) contain data meant for the master. In a data cycle, the OUT data is tnasferred from the master to the devices and the IN data is transferred from the devices to the master. The loopback word, shown in the figure, pulls the OUT data along behind it while pushing in the IN data along in front of it.

A bus cycle begins with the master passing a bit stream to the first slave. The first slave then transfers the same to the second one in the ring and at the same time data from the first slave is transferred to the master.

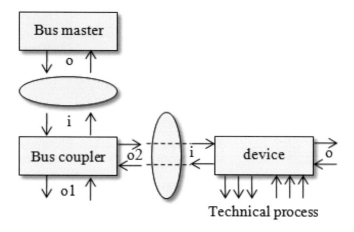

FIGURE 14.4   Components in an Interbus system. (Courtesy: Dr.-Ing. Reinhard Langmann, INTERBUS Basics. Process Control Laboratory, University of Applied Sciences, FH Dusseldorf.)

The master knows the physical position of a slave in the ring and so the slaves need not be addressed separately. The master also knows the amount of information to be delivered to each device which is ultimately included in the summation frame.

14.4   TOPOLOGY

The system comprises communication bus master, bus coupler, bus devices (slaves), and connecting the above are the local bus as well as the remote bus as shown in Figure 14.4.

The bus master, normally available as a controller board, performs the following tasks: transferring data between the host (normally an industrial PC or a PLC) and the bus device, management of the bus, and communication between the devices. The management functions include error detection, configuration, etc.

The bus coupler, also known as bus terminal device, divides the ring connection into bus segments. It functions as a slave and connects the incoming (i) and outgoing (o) interfaces. Figure 14.5 shows two types of bus couplers normally required in an Interbus system.

The bus coupler, along with the bus devices, configures the ring system. The bus devices are connected to the process signals and transfer communication signals as per requirements, i.e., either in analog or digital from. The bus coupler, under the command of bus master, can activate or deactivate either the incoming or outgoing interfaces in order to support system configuration as well as error dignostics. It is shown in Figure 14.6.

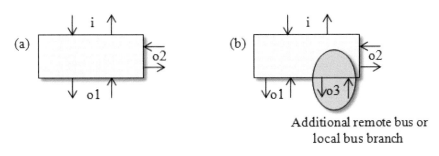

Additional remote bus or
local bus branch

FIGURE 14.5 Bus couplers: (a) standard and (b) with additional interface. (Courtesy: Dr.-Ing. Reinhard Langmann, INTERBUS Basics. Process Control Laboratory, University of Applied Sciences, FH Dusseldorf.)

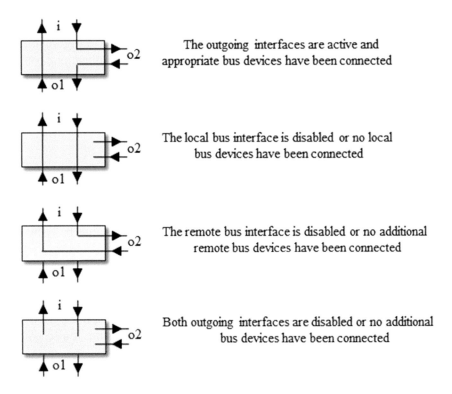

FIGURE 14.6 Techniques or activating/deactivting interfaces. (Courtesy: Dr.-Ing. Reinhard Langmann, INTERBUS Basics. Process Control Laboratory, University of Applied Sciences, FH Dusseldorf.)

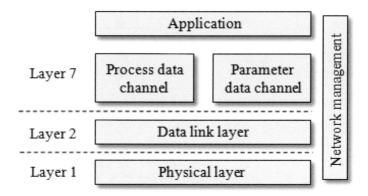

FIGURE 14.7 Interbus protocol structure. (Courtesy: Dr.-Ing. Reinhard Langmann, INTERBUS Basics. Process Control Laboratory, University of Applied Sciences, FH Dusseldorf.)

14.5   PROTOCOL STRUCTURE

The protocol structure follows the open systems interconnection (OSI) reference model and comprises layers 1, 2, and 7 and shown in Figure 14.7. Some functions of layers from 3 to 6 are included in the application layer.

Application layer helps in accessing data from the slaves (Interbus devices). Physical layer determines baud rate, data encoding while data link layer ensures data integrity. Layers 1 and 2 adhere to DIN 19 258.

14.5.1   PHYSICAL LAYER

Data transmission on the physical layer takes place at 500 kbps while the data line is scanned 16 times faster by the slaves. Data is encoded in NRZ (non-return to zero) form before transmission. There is a clock generator in each device and the individual clocks are synchronized internally by a common synchronization marker.

The format for line encoding in the physical layer is shown in Figure 14.8 and it comprises status telegram and data telegram. The status telegram transmits the status of the SL (select) signal. It is 5 bits in length and generates activity on the bus during pauses in transmission. Data telegram is 13 bits in length with 5 bits of header and 8 bits of user data. It also contains the status of the CR (control) signal.

14.5.2   DATA LINK LAYER

The summation frame telegram is shown in Figure 14.9 and corresponds to cyclic Interbus protocol in the data link layer. The methodology used is time division multiple access (TDMA) with collision-free transmission.

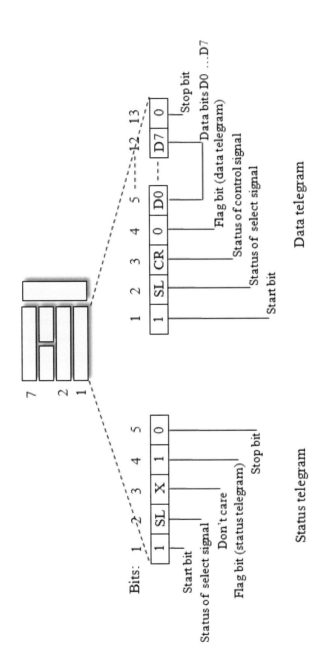

FIGURE 14.8 Line encoding format in the physical layer. (Courtesy: Dr.-Ing. Reinhard Langmann, INTERBUS Basics. Process Control Laboratory, University of Applied Sciences, FH Dusseldorf.)

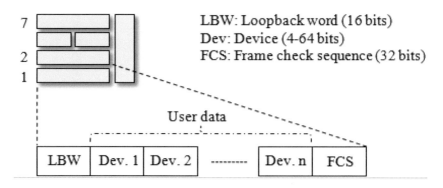

FIGURE 14.9 Summation frame telegram format in the data link layer. (Courtesy: Dr.-Ing. Reinhard Langmann, INTERBUS Basics. Process Control Laboratory, University of Applied Sciences, FH Dusseldorf.)

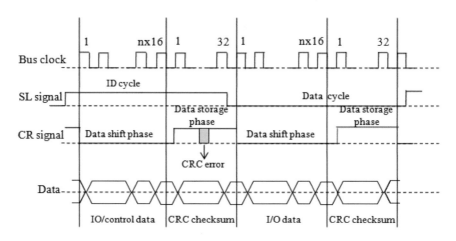

FIGURE 14.10 The different operating phases in the Interbus protocol. (Courtesy: Dr.-Ing. Reinhard Langmann, INTERBUS Basics. Process Control Laboratory, University of Applied Sciences, FH Dusseldorf.)

Each device is allocated a time in tune with its function and thus the total transition time can easily be calculated. The loop back word (LWB) is used by the master at the beginning of each cycle to ascertain the amount of data and consequently the number of shift registers in the ring. It is a 16-bit word and is used in the identification cycle (ID cycle) to detect the end of data shift phase.

Interbus protocol in the data link layer has different operating phases and the different activities are governed by SL (select) and CR (control) signals. This is shown in Figure 14.10. SL signal is characterized by ID cycle and data cycle, while CR signal by data shift phase and data storage phase.

In the ID cycle within the SL signal, the devices switch the ID registers in the Interbus ring and the master can thus identifies the devices individually. The status of the SL signal in the data cycle is made zero by the master and thus the Interbus ring is closed by the data shift registers. When the CR signal is asserted, data already transmitted during SL signal is now stored in the data storage phase or FCS. CRC checksum is employed in this phase of data storage and saving.

14.5.3 APPLICATION LAYER

Interbus devices are accessed for data via two channels: the process data channel and the parameter channel. The former is employed for process data and the latter one for parameter data. Cyclic data exchange between sensors and higher-level control systems takes place via the process data channel while the parameter channel helps accessing connection oriented message exchange. Message transmission takes place in parameter channel in client-server model and a large volume of data is exchanged between the communicating devices. Each Interbus device has a process data channel while parameter channel requirement is an optional one. Performance of application layer is dependent on how seamlessly it can handle the two data types efficiently.

Process data for a device is in the range of a few bits and comes from set point, limit switches, control signals, etc. It is almost always uniquely identified by its address and/or the sensor/actuator it represents. The collective total of I/O data in the summation frame telegram during a data cycle represents the process data.

Interbus devices with parameter channel are servo amplifiers, operating and display units, etc. and may be in the range of 10–100 bytes. Parameter data are acyclic in nature and are transmitted only when it is required between two Interbus devices.

15 ControlNet

15.1 INTRODUCTION

It is an open industrial network protocol belonging to common industrial protocol (CIP) family. It was initially supported by ControlNet International. Since 2008, the activities and management of ControlNet is being managed by the Open DeviceNet Vendors Association, Inc. (ODVA). CIP is also used in DeviceNet and EtherNet/IP.

ControlNet is included in EN 50170 and IEC 61158 standards. It is designed to be used both at the device and cell levels of industrial control systems. Applications of ControlNet include batch control systems, automotive industries and process control systems. The session layer in the ControlNet communication profile is empty, as shown in Figure 15.1.

15.2 FEATURES

ControlNet communication is a high-speed deterministic network used for transmission of time-critical applications and can be scheduled as per requirement. Common network sniffers cannot sniff into controlNet packets. In peer-to-peer communication mode, it provides real-time control and messaging services. It supports producer/consumer network model allowing data from a node to be multicast resulting in increased efficiency and higher system performance.

It offers high throughput—about 5 Mbits/sec for improved I/O performance. Other features of ControlNet include: deterministic and highly repeatable data delivery, multiple controllers controlling I/Os independent of each other belonging to the same link.

It supports both coaxial and fiber optic cables in tree, bus, or star topology. Repeaters can extend the reach of the network to distances more than 30 km. It has media layer redundancy for increased network reliability.

15.3 PRODUCER–CONSUMER MODEL

Producer-consumer model is followed in ControlNet in which all nodes access data from a single source at the same time, i.e., synchronously. Synchronism is achieved because data arrives at each node simultaneously resulting in higher performance and increased efficiency. In legacy

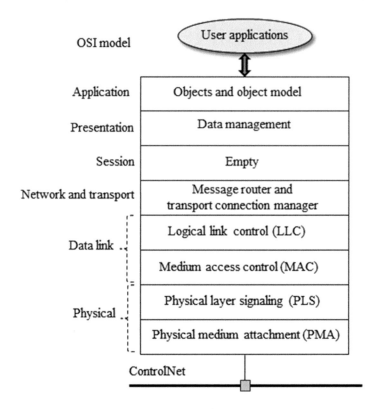

FIGURE 15.1 The ControlNet communication profile. (Courtesy: B. G. Liptak. Instrument Engineers' Handbook, Process Software Digital Network, 3rd Edition, CRC Press, Boca Raton, FL, p. 614, 2002)

source-destination model, data is received by different destinations at different times and requires multiple packets to deliver the same data to multiple nodes. It results in reduced efficiency because of extra network traffic. Also legacy system requires different networks for sending messaging and time critical I/Os. Compared to this, the producer-consumer model is fully synchronized, same network can send messaging and time critical I/Os and bandwidth is optimized for higher performance. This system supports master-slave, multimaster, peer-to-peer communication and also hybrid systems—i.e., a mix of the above three types.

15.4 CONTROLNET MEDIA

The ControlNet media consists of physical components like: connectors, taps, repeaters, bridges, cables, terminators, etc., shown in Figure 15.2.

It has one link and the link has two segments. A link is a collection of nodes having unique addresses in the range of 1–99. It also shows a

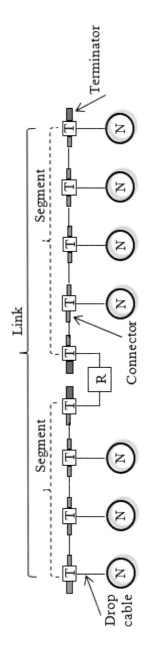

FIGURE 15.2 The ControlNet media.

repeater that replicates the signal from one segment to another. The nodes (N) connect the physical devices to the cable system via drop cables. Two terminators of value 75 Ω each are inserted on either side of a segment. Taps connect devices to the trunk cable. A tap is required for each and every node and on either side of a repeater.

15.5 CONTROLNET CONFIGURATION

All the nodes in a ControlNet configuration are governed by a set of rules. First, all the nodes in a link must abide by a set of network configuration parameters before transmitting scheduled/unscheduled data. Second, each connection originator (CO) must have specific scheduling information before it can request or transmit scheduled data. Finally, each connection target (CT) can transmit unscheduled data after power on and requested by the connection originator. Figure 15.3 shows the connection between a ControlNet configuration keeper node and CT nodes. A keeper node is a non-volatile memory device that holds network parameters and scheduling information that it distributes to other nodes.

15.6 PHYSICAL LAYER

Physical layer of ControlNet is split into two sub layers; physical medium attachment (PMA) and physical layer signaling (PLS) as shown in Figure 15.1. PMA comprises circuits that aid in delivering/accepting data to/from the bus to which the devices are attached. PLS is responsible for appropriate timing and bit representation of data generated at the devices. It also acts as an interface to the next higher layer—the data link layer.

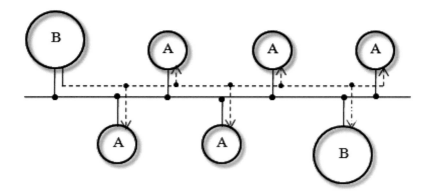

A: CT node   B: CO and keeper capable node

FIGURE 15.3 ControlNet configuration between CO and CT.

The physical media can be of three types: coaxial cable, fiber optic, and network access port or NAP. Maximum cable length is 1000 m without repeaters and the maximum number of nodes is 99. The media can support a maximum of 5 repeaters (10, if redundancy is used). The physical layer signaling uses Manchester code with a bit rate of 5 Mbits/sec.

When used with fiber optic cables, the system can have two lengths: up to 300 m for short-range systems and up to 7000 m for medium-range systems. Fiber optic type can employ both active star and active hub topologies.

NAP has eight conductors with an overall shield. It is used to establish a connection between a programming unit and a station which is already attached to ControlNet. For a point-to-point temporary connection between two nodes, NAP is used.

15.7   DATA LINK LAYER

Data link layer of ControlNet consists of two sub-layers: medium access control (MAC) and logical link control (LLC).

When a node sends data over the network, it is packed into a MAC frame. The MAC frame is shown in Figure 15.4. A single MAC frame may contain several LPackets (Link Packets). A maximum of 510 bytes of data can be sent in a single frame. The format of the LPacket is also shown in the figure, which contains a field called connection ID (CID), apart from

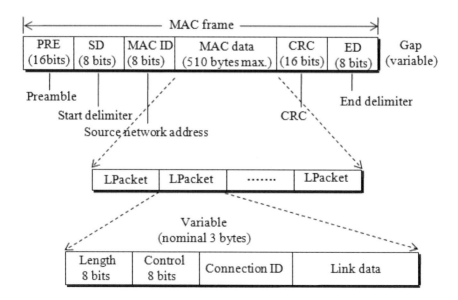

FIGURE 15.4 MAC frame and LPacket formats. (Courtesy: Rockwell Automation-Allen Bradley. An Introduction to the ControlNet Network, System Overview, Release 1.5, 1997.)

the other fields. The CID can be of two types: a 2-byte fixed CID and a 3-byte general CID.

15.8   NETWORK ACCESS MECHANISM

Access to network is based on a time slice algorithm, based in MAC. This algorithm is called concurrent time domain multiple access (CTDMA). This regulates data traffic on the network. Figure 15.5 shows ControlNet's media access mechanism. It is apparent from the figure that the network update time (NUT) is fixed to which all the nodes are synchronized. For the network, NUT is indefinitely repeated. A NUT is divided into three sections: scheduled, unscheduled, and guard band, also known as network maintenance.

Real-time data, for example, both cyclic data exchange and asynchronous data traffic are sent in the scheduled interval. In this, message delivery is fully deterministic and repeatable. Bandwidth in the scheduled portion is reserved and configured beforehand to support real-time data transfers.

Each station or node is granted only a single frame transmission in a NUT. Each station is assigned an address called its MACID. When the NUT begins, the station with the lowest MACID is granted permission to transmit a frame. Every station has an implicit register which is its own MACID. When a frame transmission is over, the MACID is automatically incremented by one. Each station compares this new ID with the contents of the implicit register. For the station for which these two match, gets the right to transmit its own frame. The scheduled transmission technique is shown in Figure 15.6.

Any station having a network address between one and SMAX (scheduled maximum) will be granted exactly one opportunity to transmit in a single NUT. Up to SMAX bandwidth is reserved in the scheduled portion of NUT. Network address zero is not allowed for a station, which is reserved for future use. Slot time is the time duration that a node or station would wait for a missing network address before initiating transmission on the network bus.

Once the scheduled interval is over, the unscheduled portion of the NUT takes over, which does not have any time critical constraints. Unscheduled transmission in NUT is neither deterministic nor repeatable. In this, the implicit token is still circulated amongst the stations on a round-robin scheme. This continues for network addresses between zero and UMAX (unscheduled maximum) until the beginning of guard band. The unscheduled transmission technique is shown in Figure 15.7.

A node may have the opportunity to transmit more than once in the unscheduled portion of NUT, but transmission is not always guaranteed. Some important characteristics of unscheduled communication are: (a) unscheduled service is from 0 to UMAX. UMAX can never be less than SMAX, (b) stations with addresses greater than SMAX and less than or

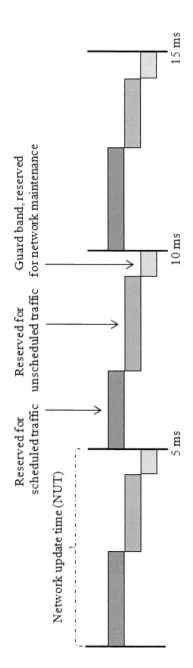

FIGURE 15.5 Media access mechanism for ControlNet. (Courtesy: Rockwell Automation-Allen Bradley. An Introduction to the ControlNet Network, System Overview, Release 1.5, 1997.)

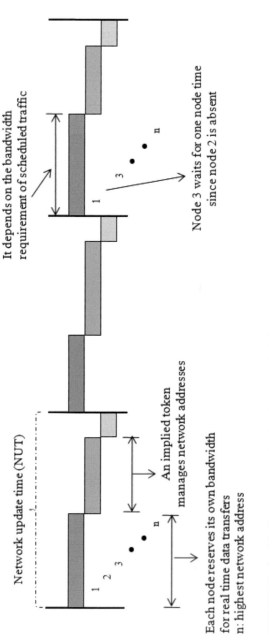

FIGURE 15.6 Media access mechanism in scheduled time in ControlNet. (Courtesy: Rockwell Automation-Allen Bradley. An Introduction to the ControlNet Network, System Overview, Release 1.5, 1997.)

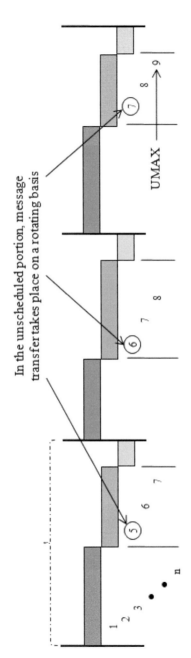

FIGURE 15.7 Media access mechanism in unscheduled time in ControlNet. (Courtesy: Rockwell Automation-Allen Bradley. An Introduction to the ControlNet Network, System Overview, Release 1.5, 1997.)

equal to UMAX may only send unscheduled messages, (c) stations with addresses less than or equal to SMAX may send both scheduled and unscheduled messages, and (d) stations having addresses greater than UMAX cannot communicate through the ControlNet network.

The scope of transmitting first in the unscheduled part of NUT passes on a rotating basis. In the first NUT, node 7 transmits first. In the second NUT, it is node 8 which transmits first irrespective of which node finished the last interval. In conclusion, it can be said that the right to transmit first in the unscheduled portion rotates by one node per NUT.

In each NUT, once the scheduled and unscheduled portions are over, guard band interval takes over. Transmission on the network is stopped and the station with the lowest network address (MACID) is granted access. This station or node is called moderator. In the guard band, this node transmits only the moderator frame, which keeps all the nodes in synchronism and also a set of parameters necessary for correct operation of the network.

15.9   NETWORK AND TRANSPORT LAYER

The three major basic modules existing on these two layers are: unconnected message manager (UCMM), message router (MR), and connection manager (CM).

UCMM can send a message without an established connection. It is used to transmit non-repetitive, non-time critical data on a single link, which are always sent in the unscheduled portion of NUT. It facilitates execution of unconnected messages. In this case, each and every data transfer is independent of others and the messages carry the descriptions of destination and source applications, which are provided by the connection manager.

As shown in Figure 15.8, message coming from the ControlNet network is passed on to message router via UCMM. The message router analyzes

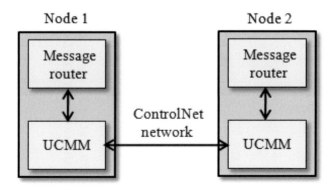

FIGURE 15.8   UCMM and message router interconnection. (Courtesy: Rockwell Automation-Allen Bradley. An Introduction to the ControlNet Network, System Overview, Release 1.5, 1997.)

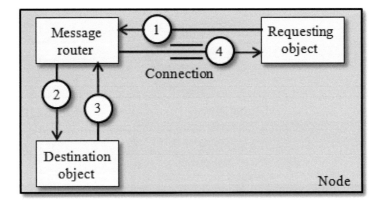

FIGURE 15.9 Functioning of a message router. (Courtesy: Rockwell Automation-Allen Bradley. An Introduction to the ControlNet Network, System Overview, Release 1.5, 1997.)

the same and sends it to the specified function or object. UCMM functions include: (a) receiving incoming UCMM messages (b) sending and receiving unconnected messages to and from (i)UCMM objects on other nodes and (ii) its local message router.

MR guides the correct dispatch of services to the addressed applications within the node. Figure 15.9 shows the functioning of message router.

MR analyzes the message to determine the service to be executed by the identified object. The message is forwarded to the destination object which in its turn sends its response to the requesting object via the MR.

Internal resources required for a connection establishment is provided by the connection manager (CM). The request for connection may come from another node via the UCMM or else an application existing on a node. Figure 15.10 shows the operation of a connection manager.

First, UCMM of the source tries to establish a connection with the UCMM of the destination with a connection request. This is routed through MR of the target to the connection manager which then allocates the requisite resources. Finally the needed connection is established.

Seven classes (classes 0–6) of connections are defined for the connection manager. They are distinguished by their features, complexities, and nature of services required of them.

15.10   PRESENTATION LAYER

It is based on IEC 1131-3 standard. Elementary and derived data types are defined by this standard. Data management is managed by this layer. It specifies the format of the data to be handled by the application layer.

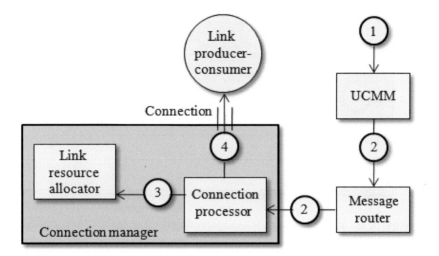

FIGURE 15.10 Functioning of a connection manager. (Courtesy: Rockwell Automation-Allen Bradley. An Introduction to the ControlNet Network, System Overview, Release 1.5, 1997.)

15.11 APPLICATION LAYER

The ControlNet network application layer is based on object modeling. Object modeling refers to organizing related data and procedures into one entity: the *object*. An object can be thought of as a collection of related *services* and *attributes*. Application objects share their resources and information by sending messages. Messages can be sent between two objects of a node or across the ControlNet network. The message paths of such objects are shown in Figure 15.11.

A standard is an object library which contains all the details of the objects used in a station. Sometimes vendor-specific objects are included in the standard by incorporating their device profiles so that devices from different manufacturers can be used interchangeably.

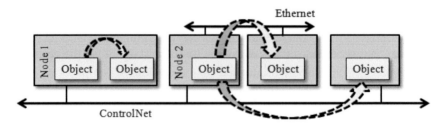

Messages are shown by curved dashed arrows

FIGURE 15.11 Message paths for application objects. (Courtsey: www.dia.uniroma3.it/autom/Reti_e_Sistemi_Automazione/PDF/ControlNetDetails.pdf)

16 Common Industrial Protocol

16.1 INTRODUCTION

Initially, networks used in manufacturing and process control industries were optimized for specific applications. Thus, although it met the specific networking needs of a particular industry satisfactorily, it failed to meet the desired level of performance with regard to efficiency and reliability in other industries. Network users were thus left with choosing the one that performed well and suit their purpose. In general, these disparate networks were not fully interoperable with each other and as such users using more than one network in a manufacturing unit were left with isolated islands of automation in which a network cannot communicate with another fully because of incompatibility issues.

Today's corporate managers seek to have a seamless connectivity any time anywhere from the shop floor to top management level—encompassing the whole gamut of automation levels. With exponential rise in Internet technologies, automation industry users seek to have an *open* system so that different networks can seamlessly be connected.

Common Industrial Protocol (CIP™) is an open, connection-based communications protocol that circumvents the above problems of seamless connectivity across different protocols—the Open DeviceNet Vendors Association, Inc. (ODVA) media independent network protocol. CIP encompasses a comprehensive suite of messages and services for varied industrial applications that include control, safety, energy, synchronization, motion, information, and network management. It allows users to integrate the above functions (applications) with enterprise level Ethernet networks and the Internet. CIP provides a truly unified communication architecture at the upper layers of the seven layer open systems interconnection (OSI) protocol and provides interoperability and interchangeability that is required for an open system. The extensions of CIP for critical applications are: CIP Safety, CIP Motion, and CIP Sync.

The Open DeviceNet Vendors Association, Inc. (ODVA), founded in 1995, is a global organization of world's leading automation companies and its main objective is to have an open and interoperable communication technology. ODVA promotes the adoption of CIP in manufacturing automation systems and the network adaptations of CIP—EtherNet/IP™,

DeviceNet™, CompoNet™, and ControlNet™. ODVA manages, develops, and distributes the specifications of these four networks with a common structure to help ensure consistency, reliability, and accuracy. Figure 16.1 shows the organization of the library of the four networks.

16.2   FEATURES

The main features of CIP are as follows: (a) It is a truly open, media-independent protocol supported by hundreds of vendors around the world. (b) It provides a unified communication architecture throughout the manufacturing enterprise. (c) It encompasses a suite of messages/services for collection of manufacturing automation applications like control, safety, energy, synchronization, motion, configuration, information, and network management. (d) It protects current investments while going in for any expansion. (e) It provides seamless bridging and routing without incurring additional cost. (f) It delivers a single media-independent protocol for the four network adaptations of CIP, viz., EtherNet/IP, ControlNet, deviceNet, and CompoNet.

16.3   FAMILY OF CIP NETWORKS

As already stated, CIP adapted four protocols are: EtherNet/IP, DeviceNet, ControlNet, and CompoNet.

Network tools in EtherNet/IP deploy standard Ethernet technology and at the same time enables Internet and enterprise connectivity for industrial automation applications to be made available any time anywhere. Topology options available with EtherNet/IP are the conventional star with standard Ethernet infrastructure devices and device-level ring (DLR) with EtherNet/IP devices. DeviceNet is CIP on CAN technology. It is a low-cost technology employed for simple devices. It uses a trunkline or dropline topology. DC power required for the network is run on the network cable for ease of installation. Quickconnect function is an option available with both EtherNet/IP and DeviceNet for connecting devices quickly on the network while the network is running. ControlNet is used for high speed, deterministic transport of time-critical I/Os, and peer-to-peer interlocks. Topology options used are star, tree, trunkline, or dropline. CompoNet is used for very quick transportation of small packets of data between controllers, sensors, and actuators and has a simple cabling scheme.

16.4   OBJECT MODELING

Every network device (or node) in a CIP protocol is modeled as a series of objects. An object is an abstract representation of a particular component within a product or simply a grouping of related data values in a device. An object modeling is used to describe (a) the behavior of a node (b) the

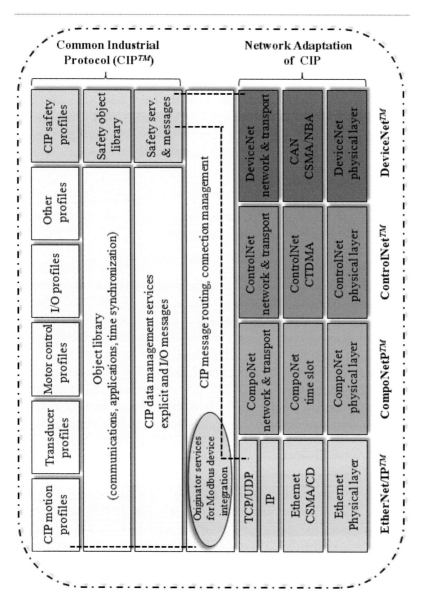

FIGURE 16.1 Organization of the library of CIP adapted four networks. (Courtesy: The Common Industrial Protocol (CIP) and the family of CIP Networks, Pub No: PUB 00123R1, www.odva.org)

suite of available communication services and (c) the manner by which information within CIP products is accessed and exchanged. CIP objects are classified into classes, instances, and attributes. A class represents a set of objects having identical type of system component. An object instance is the actual representation of an object within a class. Each instance of a class has identical attributes.

Every CIP device is required to be identified in the network—the identity object does this job. Attributes for an identity object include vendor ID, date of manufacture, device serial number, and any other relevant identity data.

Each node (device) in a CIP network is identified by a node address. This address is the MAC ID for DeviceNet, ControlNet, and CompoNet networks while it is the IP address for EtherNet/IP. Again there are class identifier, instance identifier, and attribute identifier for each node.

16.4.1 OBJECT TYPES

In CIP, objects can be of different types like: required objects, optional objects, application objects, and vendor-specific objects.

16.4.1.1 Required Objects

Required objects are: identity object, message router object, network-specific link object, and connection object. The identity object is the information that identifies a CIP node. External devices access a CIP device through the message router object. Network-specific link object provides information about the specific link required to implement the CIP device. The specific link may be EtherNet/IP, DeviceNet, ControlNet, etc. Characteristics of connection for a CIP device are maintained in the connection object. An instance of the connection object is generated for every connection. The instance identifies the connection as explicit or implicit.

16.4.1.2 Optional Objects

Optional objects are: assembly object, parameter object. Instances of an assembly object organize data which is exchanged with external devices. There can be input or output assembly instances. The former organizes data that is transferred to external devices while for the latter, data is transferred from external devices. A parameter object provides configuration parameters of a CIP device to external device.

16.4.1.3 Application Objects

There are many devices residing on top of the application layer of CIP. The devices may be pneumatic controllers, AC drives, motion control, drive systems etc. In the general case, more than one device for each device type exists in an industrial set up. An electrical motor or a drive system will have attributes such as current, frequency, size, etc.

The application layer objects are predefined for the majority of common device types. Same device types have identical application objects. The series of application objects for a particular device is called the profile of the device. Device profile specifies configuration options and I/O data formats. A user can thus switch over from one device vendor to another one because the profiles of the devices belonging to the two vendors are identical. Thus, interoperability among various device types within the CIP network is achieved.

16.4.1.4   Vendor-Specific Objects

Vendor-specific objects are not found in the profiles of a device and are included by the vendor of the device. They can be accessed by the CIP protocol, if required. Data corresponding to vendor-specific objects may be organized as cyclic, polled, or change-of-state.

16.4.2   A TYPICAL DEVICE OBJECT MODEL

A typical device object model is shown in Figure 16.2. It consists of various types of objects, although in a given situation, only a subset of these device objects is used. As already mentioned, out of the four types of objects, the *required object* type must be used for any implementation.

16.5   CIP MESSAGING PROTOCOLS

CIP is mostly a connection-oriented protocol. When application objects are connected by a path, a connection ID (termed CID) is assigned between the connecting objects. In case of bidirectional data exchange, two CIDs are needed to be assigned. An unconnected message manager (UCMM) function manages connection between devices which are yet to be connected. UCMM is responsible for processing and establishing a connection for explicit requests and responses. Figure 16.3 shows a method of connection establishment in CIP networks.

16.5.1   IMPLICIT AND EXPLICIT MESSAGING CONNECTIONS

In a CIP network, there are two types of connections: I/O (implicit) connection and explicit messaging connection.

I/O messaging can be unicast or multicast in nature and a dedicated path is established between a producing application and one or more consuming applications. Since meaning of data is implied in the connection ID (CID) itself, it is also called implicit messaging. A multicast I/O connection in a CIP network is shown in Figure 16.4.

Explicit messages are request-response type communications and the messages are point-to-point. These messages are called explicit because the data explicitly tells what service and object is being requested. Figure 16.5 shows an explicit messaging connection.

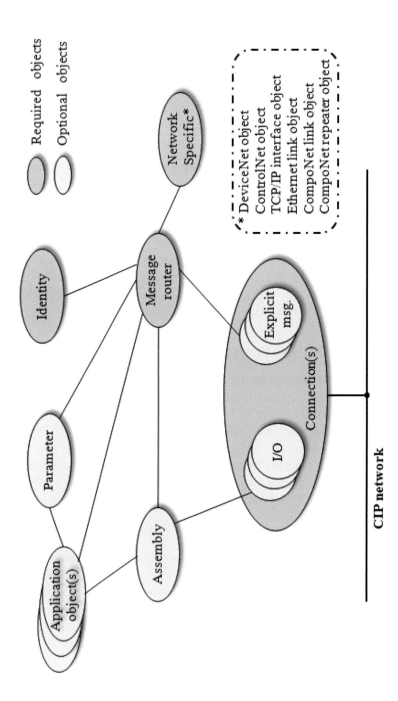

FIGURE 16.2 A typical device object model. (Courtesy: The Common Industrial Protocol (CIP) and the family of CIP Networks, Pub No: PUB 00123R1, www.odva.org)

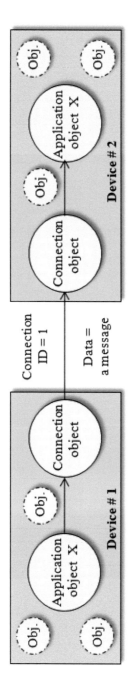

FIGURE 16.3 CIP connection establishments. (Courtesy: The Common Industrial Protocol (CIP) and the family of CIP Networks, Pub No: PUB 00123R1, www.odva.org)

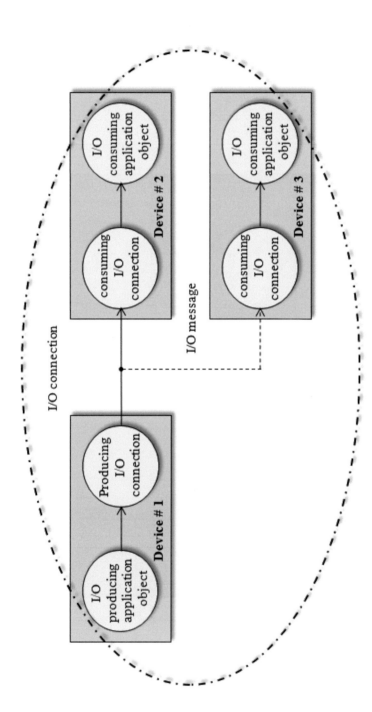

FIGURE 16.4 A multicast I/O connection. (Courtesy: The Common Industrial Protocol (CIP) and the family of CIP Networks, Pub No: PUB 00123R1, www.odva.org)

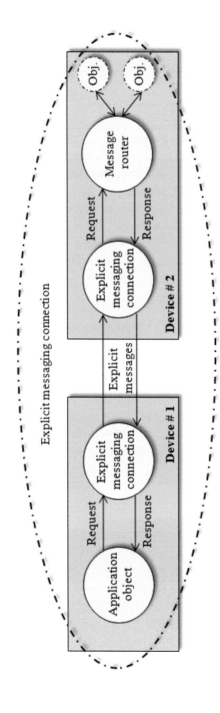

FIGURE 16.5 An explicit messaging connection. (Courtesy: The Common Industrial Protocol (CIP) and the family of CIP Networks, Pub No: PUB 00123R1, www.odva.org)

16.6   CIP APPLICATION LAYER ENHANCEMENTS

The application layer of CIP is enhanced, depending on the application needs, in areas of synchronization, motion, and safety. These are: CIP Sync™ (synchronization), CIP Motion™ (real-time, deterministic, multi-axis motion control), and CIP Safety™ (functional safety).

16.6.1   CIP Sync

It is based on IEEE 1588 standard—precision clock synchronization proto-col for networked measurement and control systems. It is applied in distrib-uted motion control, sequencing mechanism that record events, etc. Using the standard, transmission time latencies as well as infrastructural delays can be limited to within hundreds of nanoseconds accuracy.

CIP sync is an extension of EtherNet/IP in the real-time domain. It per-forms nicely without any segmentation with other parts of the communica-tion system that is less demanding on real-time performance.

16.6.2   CIP Motion

It manages real-time motion control with the help of CIP sync. CIP motion is achieved by including time as a part of motion information. The distributed nature of the clock used in CIP motion ensures both data and time stamp being delivered in the packet to the network. Hence, the motion controller and the associated drives receive the time stamp at the same time resulting in coordinated multi-axis motion. This method relieves the network from a very rigid and rigorous time data delivery schedule. The motion execution remains unaffected even if there is a slight fluctuation in data delivery.

16.6.3   CIP Safety

It allows mixing of safety devices and standard devices on the same net-work—this is because safety functionality is incorporated in each device and not in the network infrastructure. This gives the network user the free-dom to use a device either with safety or without it. CIP safety is defined in a well-defined layer and it is scalable as well as network independent. Safety switches, safety PLCs, etc. can be used up to Safety Integrity Level 3 (SIL 3) as per IEC 61508 standards.

16.7   NETWORK INDEPENDENCE OF CIP SAFETY

CIP safety protocol was designed to be network independent, unlike other networks for which safety protocols require different profiles. CIP safety is a highly integrated safety services which leverage the underlying

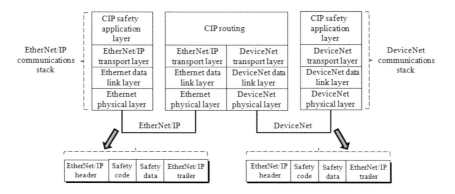

FIGURE 16.6   A safety network spanning across EtherNet/IP and DeviceNet. (Courtesy: CIP safety brings network independence to EtherNet/IP, D. Vasko, www.odva.org)

communication stacks of the networks that support CIP protocol, viz., EtherNet/IP, DeviceNet, CompoNet and ControlNet. Thus, CIP safety is an open protocol that seamlessly bridges networks that support CIP.

The safety protocol detects communication errors such as message error, data loss, delay in transmission, etc. The safety code is executed in a high integrity section of the safety enabled devices. When an error occurs, the device goes into safety mode.

For an EtherNet/IP safety device, the safety packets are encapsulated in the EtherNet/IP data frame that comes out from the physical layer. It is alike for a DeviceNet data frame. Because of network independence nature of safety protocol in CIP, a safety connection can span across different networks. Since data integrity depends on safety code and independent of underlying communication layers, a safety connection can start on EtherNet/IP and terminate on DeviceNet or vice versa. This is shown in Figure 16.6.

16.8   BENEFITS

CIP is an open and media (network)-independent communication protocol that sits at the top of several networks. A number of advantages accrue by implementing CIP across different networks. It implements a producer-consumer approach and thus enjoys efficient bandwidth utilization. It provides a network-independent architecture that allows seamless communication from shop floor level up to the enterprise level. It has a comprehensive suite of messages and services for manufacturing automation sector and provides different functionalities like synchronization, safety, control, configuration and motion. The device profiles in CIP provide a common application interface. A seamless bridging and routing enables multiple topology options without reprogramming or reconfiguring intermediate devices. CIP is implemented across different networks and supported by multi vendors.

17 Ethernet and Ethernet/IP

17.1 INTRODUCTION TO ETHERNET

Ethernet was developed by Xerox's Palo Alto Research Center (PARC) in 1976. It is the most popular physical layer local area network (LAN) technology in use today. It is standardized in IEEE 802.3 which governs the rules for configuring an Ethernet network and specifies the manner by which one device communicates with another.

Ethernet is the global standard that connects multiple computers, machines, devices to join each other with wires and cables to form a network. It has become very popular because it strikes a very good balance between speed, cost, and ease of installation.

Ethernet, when it was introduced more than four decades ago, was mainly used in data communication purposes. Demands for real-time data transfer came much later when Ethernet technology was applied in industrial environments. Initially, it was connected in a network with a LAN configuration.

17.1.1 DIFFERENT GENERATIONS OF ETHERNET

Since it was introduced in the market more than four decades back, Ethernet has gone through four generations: Standard Ethernet (10 Mbps), Fast Ethernet (100 Mbps), Gigabit Ethernet (1 Gbps) and Ten Gigabit Ethernet (10 Gbps). The standard Ethernet is also known by first or traditional Ethernet.

Standard Ethernet is categorized into four versions: 10Base5, 10Base2, 10Base-T, and 10Base-F.

Fast Ethernet (IEEE 802.3u) was introduced to meet the demands of higher speed required at that time. It required very small changes in cabling to switch over from standard to fast Ethernet version. It provides higher throughput for video, multimedia, etc. and has stronger error detection and correction ability. Fast Ethernet is available in three versions: 100BASE-TX, 100BASE-FX, and 100 BASE-T4. Gigabit Ethernet was developed for even faster communication like in multimedia systems and voice over IP (VoIP).

17.1.2 FRAME FORMAT AND LENGTH

The frame format of Ethernet along with the minimum and maximum payload lengths are shown in Figure 17.1. The format contains seven

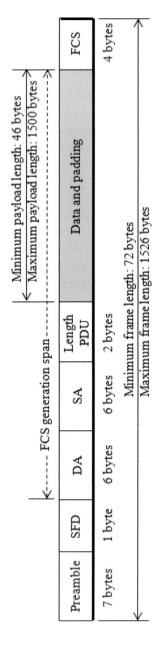

FIGURE 17.1 The Ethernet frame.

fields. The first field is a seven byte preamble of alternating 1's and 0's. It alerts the receiver that a frame is coming and that synchronizes it with the receiver. SFD or start frame delimiter (10101011) indicates the beginning of the frame. The last two bits 11 of SFD alert that the next field is the DA field. DA and SA indicate the address of destination and source stations, respectively. The length PDU is either a type field or length field. The type field indicates the protocol of the upper layer while the length indicates the number of bytes in the data field. The data field is encapsulated from the upper layer with a minimum and maximum of 46 and 1500 bytes, respectively. The last field is frame check sequence (FCS) which is CRC-32 and is used for error detection.

17.2 INDUSTRIAL ETHERNET

Ethernet was initially introduced for data communication in the office environment. A switched Ethernet architecture is the most judicious choice when applying Ethernet in an industrial setup. Several users can send information over the network at the same time via the switches without slowing each other down.

Industrial Ethernet differs from standard Ethernet in traffic prioritization, robust equipment, hardware, and industrial-grade components although both use the same traditional Ethernet technology. Industrial Ethernet will have to withstand harsh industrial environments in the form of high temperature, hostile environment, dust, vibration, and shock. They also differ in power supply requirements and need a fault tolerant network. Industrial Ethernet applies Ethernet standards developed for data communications to industrial control networks and can handle data collisions in a better manner at the shop floor level. Cables and connectors for such networks are more robust in nature.

Normally, producer-consumer communication is the norm in industries—i.e., data (or information) produced by a device is consumed by many devices in parallel. Thus, it is multicast in nature. A traditional Ethernet network relies more on efficient utilization of bandwidth while its industrial cousin focuses more on synchronous data transmission and access. Thus, to prioritize multicast data transmission, it needs intelligence and quality of service (QoS) features. An industrial Ethernet network should have features like reliability, network security, and determinism.

17.3 EVOLUTION OF REAL-TIME FIELDBUSES

Ethernet and TCP/IP protocols were found to be good enough to control manufacturing operations on the shop floor, but could not control communications inside machines and equipment. These two protocols failed to satisfy real-time deterministic communication from machine control to actuators and sensors.

Use of Ethernet as a fieldbus grew at a frenetic pace over the years primarily because of easy availability of Ethernet components as well as their low cost. Network interface cards (NICs) and CAT5 Ethernet cables were standardized and their cost fell to a great extent which is well below the cost of a proprietary fieldbus-based system.

This thus led to developments of many industrial Ethernet protocols which could use the same hardware components as used in Ethernet-based networks and at the same time would be real time and deterministic. Some among the leading control manufacturers used Ethernet components in their protocols but made them proprietary. Users of these protocols were thus tied to these manufacturers for any future needs and expansions. This led to several open standard real-time Ethernet fieldbuses and the most sought after fieldbuses are: EtherCAT, Ethernet Powerlink, EtherNet/IP, SERCOS III, and PROFINET IRT. What makes these fieldbuses stand out from others are interoperability, openness, conformity, easy availability of components, ease of implementation, ease of future expansions, etc.

17.4   REAL-TIME DETERMINISTIC FIELDBUS REALIZATION

There are three different approaches that allow a standard to realize real-time determinism. These are described in the following list:

- **Standard software/standard Ethernet:** Fieldbuses under this category rely on TCP/IP layers with real-time determinism embedded in the top layer. This has a limited real-time realization.
- **Open software/standard Ethernet:** In this case, new standard protocols are implemented on top of standard Ethernet layers. Such a standard include proprietary software controller corresponding to open systems interconnection (OSI) layers 3 and 4 to reserve time on the network. This is done to eliminate latency.
- **Standard software/modified Ethernet:** In this case, combination of a new protocol and hardware ensures determinism. The software is available in the public domain. The hardware can have a special switch or an ASIC embedded into a slave device.

As industrial production becomes more and more automated, hard real time is becoming more demanding. Different processes and their response times are shown in Figure 17.2.

Time vs. processes

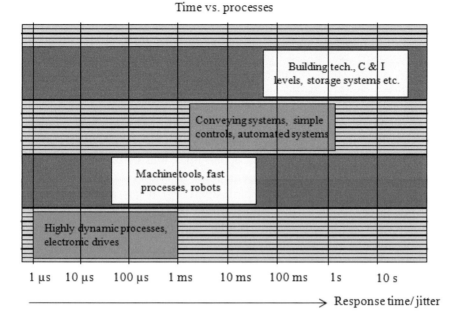

Real time classes and application areas (IAONA classification)

FIGURE 17.2 Different processes and their response times.

17.5 INTRODUCTION TO ETHERNET/IP

With industrial automation, came machine controls that depended heavily on proprietary networks and thus suffered from interoperability issues. Use of appropriate real-time Ethernet fieldbus for machine control operations gave a definite edge to machine builders vis-à-vis cost and performance.

At present, there are many open standard real-time Ethernet fieldbuses offering price and performance advantages. Out of the many, EtherCAT, Ethernet Powerlink, EtherNet/IP, SERCOS III, and PROFINET IRT are the dominant fieldbus technologies.

EtherNet/IP (Ethernet Industrial Protocol) was developed by Allen Bradley (Rockwell Automation) and ODVA (Open DeviceNet Vendors Association) in 2000. EtherNet/IP is essentially a port of the common industrial protocol (CIP) application protocol.

EtherNet/IP is an application layer protocol on top of TCP/IP and uses the same physical, data link, network and transport layers. It also uses CIP over TCP/IP. It is an open, vendor neutral, communication protocol for a wide range of manufacturing applications which can be integrated with the enterprise.

17.6   FEATURES

The main features of EtherNet/IP are as follows: (a) It offers producer-consumer services and a very efficient peer-to-peer communication between slaves. (b) It is deterministic in nature with limited real-time capability. It is time based and not cycle based. It implies that control commands should reach field stations in time. (c) Real-time delivery is based on UDP, prioritization (QoS), and IEEE 1588. (d) It is compatible with OPC Internet protocols like HTTP, SNMP, FTP, and DHCP. (e) Flexible installation options like copper, fiber, and wireless. (f) To realize real-time deterministic communication, it follows standard software/standard Ethernet architectural philosophy. (g) It has a response time of 1 ms, jitter less than 1 ms and data rate 100 Mbps. (h) Both half and full duplex operations possible. (i) UDP and TCP are respectively used for real and non-real time communication. (j) Cyclic with short update periods (controller to drive and vice versa and controller to controller) as well as long update periods (controller to output module) are supported. (k) Both unicast and multicast addressing supported. (l) It uses a CSMA/CD media access control (MAC) model that determines how networked devices share a common bus.

17.7   ETHERNET/IP LAYER MODEL

The EtherNet/IP layer model is shown in Figure 17.3. As with all CIP networks, EtherNet/IP implements CIP technology from session layer upwards and adapts it to EtherNet/IP technology at the transport layer (layer 4),

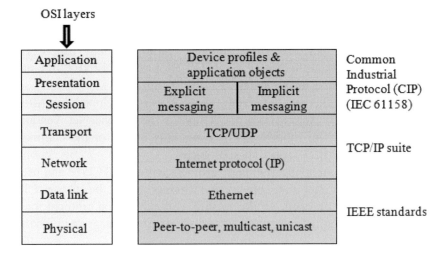

FIGURE 17.3   The EtherNet/IP layer model.

where UDP and TCP are used for time critical and non-time critical data transmission, respectively. TCP/IP encapsulation allows a node on the network to embed a message coming from above (i.e., CIP portion) as the data part in an Ethernet message.

At the data link layer (layer 2 of OSI), EtherNet/IP employs CSMA/CD media access control (MAC) technique to determine how networked devices share a common bus and how they detect and respond to collisions. Exchange of time critical data is based on producer-consumer model in which data produced by a producer can be consumed by one or more consumers. This is supported by both CIP and EtherNet/IP multicast service.

17.8 PRODUCER-CONSUMER MODEL

EtherNet/IP is based on consumer-producer model. This is shown in Figure 17.4. For such a model, data is not tied to any explicit source and destination addresses. EtherNet/IP which is an application-based protocol, meet the requirements of industrial automation architectures.

It provides very efficient slave-to-slave communication. Each consumer requires a filter because of broadcast nature of communication.

17.9 INFRASTRUCTURE FOR ETHERNET/IP APPLICATIONS

EtherNet/IP supports both time critical and non-critical data and its infrastructure can be both passive and active in nature. Fieldbuses are normally a cable-based system and inherently a passive network. Contrast to this, Ethernet-based systems including EtherNet/IP, have an active

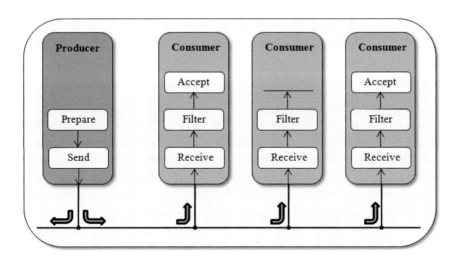

FIGURE 17.4 The producer-consumer model of EtherNet/IP.

infrastructure. Here, layers 2 and 3 are connected by switches. Switches in layer 2 pass all protocols, while those in layer 3 act as routers, allowing only designated protocols to pass through certain locations.

EtherNet/IP has to satisfy end-to-end response time requirements. Network delays are, in general, less than delays encountered in nodes (end devices). Since a controller has to satisfy both time critical and non-time critical data, they are equipped with two Ethernet interfaces. There is no need to deploy two separate physical networks to achieve the above. Instead, switches that support VLAN (virtual LAN) are used. Such VLANs create two virtual networks on the same wire.

When an EtherNet/IP network is integrated with higher-level enterprise network, infrastructural support must ensure that time critical data must not go into the enterprise part of the network. Likewise, traffic, which is meant only for enterprise network, must not go to the EtherNet/IP network to pull down its performance level.

17.10   CHARACTERIZATION OF ETHERNET/IP TRAFFIC

In EtherNet/IP, explicit messaging corresponds to non-time critical data exchange and also data generated during programming, configuration, and diagnosis of devices. These are characterized by low data rate and does not normally influence network performance. Such traffic can be unicast or multicast type. Unicast traffic consists of TCP/IP packets, while multicast traffic may be ARP, BootP, DHCP, etc. and supported by IP packets.

Implicit messaging generally consists of traffic generated during time critical data exchange and consists of UDP/IP unicast and multicast packets. Such messaging include (a) when a programmable controller produces data for other programmable controllers, and (b) when a remote I/O device produces I/O data and status for consumption by one or more programmable controllers.

Implicit messaging traffic generated by devices is of the order of tens of thousands of packets per second. Such traffic is evenly distributed between UDP/IP unicast and multicast packets—a packet is typically 120 bytes in length. Now, handling of UDP/IP unicast and multicast traffic are not identical. While the former does not require any special features in EtherNet/IP network infrastructure, the latter does require the same.

17.11   COEXISTENCE OF ETHERNET/IP WITH CIP

Figure 17.5 shows how CIP sits on top of Ethernet-based protocols like EtherNet/IP. In the transport layer, EtherNet/IP is based on TCP and UDP for time critical and non-time critical message transmissions,

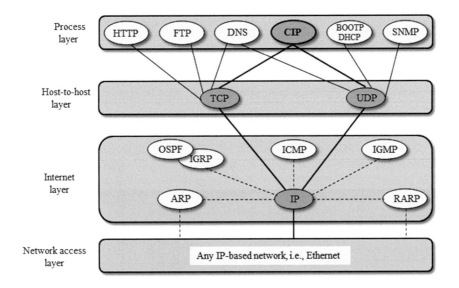

FIGURE 17.5   Coexistence of EtherNet/IP with CIP. (Courtesy: The Common Industrial Protocol (CIP) and the family of CIP Networks, Pub No: PUB 00123R1, www.odva.org)

respectively. CIP sits at the process layer along with other existing services like HTTP, FTP, etc. Thus, CIP is just another service like the other ones.

17.12   MESSAGE PRIORITIZATION IN EHERNET/IP

The maximum payload of an Ethernet frame is 1500 bytes. It takes about 122 μs for the frame to complete its transmission. This time is quite high and is likely to congest the transmission medium such that any time critical data will fail to achieve determinism. A way out to achieve determinism is to prioritize such time critical data to have the right of way as far as transmission of data is concerned. This technique is called message prioritization, also called QoS.

EtherNet/IP defines two QoS: Differentiated Services (Diffserv) and IEEE 802.1Q tagged frames. Such frames are supported by many switches. QoS allows Ethernet frames with higher priority to jump queue. Such higher priority messages will be transmitted ahead of standard EtherNet/IP or other Ethernet frames. The higher priority packets are marked with particular numbers. Routers and switches are able to differentiate between high priority packets from normal ones using QoS. The scheme is shown in Figure 17.6.

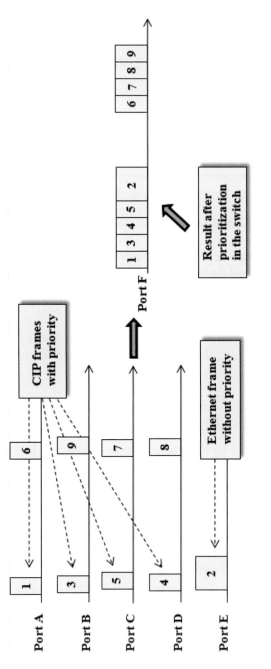

FIGURE 17.6 Prioritization of an Ethernet frame. (Courtesy: The Common Industrial Protocol (CIP) and the family of CIP Networks, Pub No: PUB 00123R1, www.odva.org)

CIP frames with priority are marked 1, 3, 4, and 5 coming from ports as shown. An Ethernet frame without priority is marked 2 coming from port E. The prioritized frames are put into the output frame F in the order 1, 3, 4, 5. This is in conformance with their relative appearance time. This is followed by frame 2 and successively the frames 6, 7, 8, and 9—which are all non-prioritized ones.

18 EtherCAT

18.1 INTRODUCTION

EtherCAT (Ethernet Control Automation Technology), developed by Beckhoff Automation GmbH, Germany in 2003, is an open protocol based on Ethernet physical layer and is now managed by EtherCAT Technology Group (ETG). It is standardized in IEC 61158 and uses standard Ethernet frame IEEE 802.3. EtherCAT uses a summation frame, *processing on the fly*, and an open software/modified Ethernet approach for faster data transfer.

EtherCAT is based on a master-slave principle in which the master sends an EtherCAT frame encapsulated in an Ethernet frame that passes through each node (slave). Each slave can read data relevant to that slave and can write (add) data to the telegram before it moves to the next slave. This way the telegram makes a complete cycle through all the connected nodes before returning to the master. Reading from or writing into each slave is enabled by a special application specific integrated circuit (ASIC) embedded into each slave. This hardware-based approach reduces latency in the nodes (slaves) and chances of collision are also eliminated.

Since control and automation applications involve a few data bytes, several nodes and possibly all the nodes in a network are combined together to form an EtherCAT frame. It may so happen that several such EtherCAT frames are put together and the whole is encapsulated to form the Ethernet frame.

A distributed clock, started by the master, is used for synchronization and conforms to IEEE 1588 time protocol standard.

EtherCAT is well suited for multi-axis motion control systems which require hard real-time data transfers between devices connected in the network.

Two versions of EtherCAT—EtherCAT P and EtherCAT G were introduced in 2016 and 2019, respectively. In the former, communication data and power for the system are transported via the same cable. EtherCAT G is applied for transferring data at Gigabyte level.

18.2 FEATURES

EtherCAT is by far the fastest industrial Ethernet technology. It is *open* in nature and thus interoperable with others. It uses summation frame for data transfer with a synchronization accuracy in the range of nanosecond.

EtherCAT architecture reduces CPU load by around 25–30 percent, compared to other bus systems, assuming same bus cycle. It does not use hubs and switches. It can have line, tree, star topologies or any combinations of them. EtherCAT has unique diagnostic and configuration capabilities, which makes it less costly and downtime is also reduced to a minimum. Any disturbance in the system can be detected down to the exact place of occurrence, thereby immediately discarding the affected nodes and network segments. EtherCAT can assign node addresses automatically, thereby eliminating the need for manual configuration. It can transport transmission control protocol/Internet protocol (TCP/IP) data along with control data because of large bandwidth present.

In a network based on EtherCAT, different cable types can be used in different segments that best suit them. In EtherCAT P version, data and power can be sent via the same cable. For consecutive node distances in excess of 100 m, fiber optic cables are used. In a particular EtherCAT segment, up to 65,535 nodes can be connected and thus network expansion is virtually unlimited.

EtherCAT uses a flexible bus cycle—both short and long versions. The former is used for refreshing data on the drives while the latter is used for sampling I/Os.

18.3   ETHERNET–BASED WORKING OF EtherCAT

EtherCAT is based on Ethernet physical layer and utilizes the advantageous features of the same. EtherCAT caters to control automation field in which each node (slave) has a few bytes of data to be transmitted but sometimes needing hard real-time communication.

In the Ethernet frame, the shortest frame length is 84 bytes. Out of 84 bytes, only 46 bytes are process data. If a node has 5 bytes of process data, then only 5/84 or 5.95 percent of datagram length is utilized. To overcome such shortcomings, several nodes or for that matter, all the nodes in a segment are encapsulated in one EtherCAT frame which then acts as one datagram of Ethernet frame. Since process data from all the nodes are accommodated in a single EtherCAT frame, it is called a *summation frame*. By doing so, overhead of the EtherCAT frame is drastically reduced. This effectively puts EtherCAT's data rate at over 90 percent.

The EtherCAT master sends the datagram that passes through each node. Each slave (field device) reads the data addressed to it on the fly and at the same time inserts its own data into the datagram. This way it proceeds until the datagram reaches the last device. The last node detects an open port and sends it back to the master by using Ethernet's full duplex facility. This keeps the delay to a minimum ensuring real-time communication. This is shown in Figure 18.1.

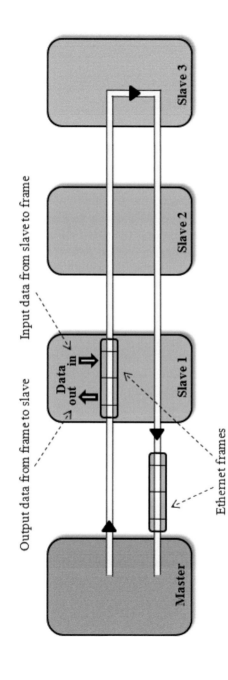

FIGURE 18.1 Master slave communication in EtherCAT. (Courtesy: Specialized Vision Sensor for Positioning, FZM1, EtherCAT Communication Manual, Cat. No. Q179-E1-01, OMRON.)

The control unit uses a standard Ethernet media access controller (MAC) without any additional controller. Thus, the EtherCAT master can be implemented on any hardware platform having an Ethernet port. An EtherCAT slave controller (ESC), residing in each slave or device, processes the frame on the fly entirely in hardware. This results in vastly improved network performance.

18.4 ETHERCAT PROTOCOL

EtherCAT frame is encapsulated in a standard Ethernet frame and is shown in Figure 18.2. The EtherCAT frame consists of a frame header and one or more datagrams. A datagram consists of its own header, process data area, and a working counter (WKC). The working counter is incremented by the devices which have been addressed by the EtherCAT command and have exchanged data already. The EtherCAT frame is identified with the identifier (0x88A4) and consists of short process data. TCP/IP and UDP/IP protocol stacks are normally not accommodated in an EtherCAT frame. If needed, they can be inserted using tunneling through a mailbox. This, however, does not impact real-time data transfer. During startup, the control unit of the master configures and maps process data on the connected devices.

The frame header in an EtherCAT frame indicates the kind of actions the master is going to execute: (a) read, write, or read/write both, (b) access a particular field device by means of direct addressing or access multiple field devices through logical addressing—also called implicit addressing.

In the EtherCAT header of a datagram, several commands are inserted, each of which addresses an individual device or memory area. The EtherCAT commands can be coded either by a special EtherType or UDP/IP. In the former case, the datagrams are not forwarded via routers. In the latter case, overhead is more because routing takes place via IP and is used for non-time critical applications.

18.5 SYNCHRONIZATION

In some specific applications, synchronization of operations of several processes or devices is a must. Contrast to complete synchronization of all the devices at the same time, the nodes or devices can be synchronized based on distributed clocks.

In motion control, multiple servo axes require highly precise synchronized operations for exact coordinated motion to be carried out. Such actions can only be achieved by distributed clocks applied to them instead of a single complete synchronized clock which suffers from communication error and jitter. The clock is based on IEEE 1588 precision time protocol standard.

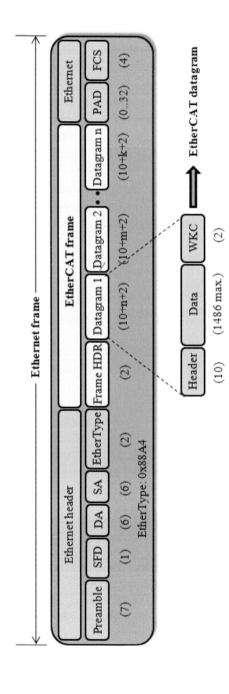

FIGURE 18.2 EtherCAT protocol stack encapsulated in an Ethernet frame. (Courtesy: EtherCAT-The EtherNet Fieldbus, EtherCAT Technology Group. www.ethercat.org)

Hardware-based calibration is done for synchronization based on distributed clocks. Because of spatial nature of distribution of devices in a process, the synchronized clock that is sent from the master will be received by devices with a delay—this is more so for devices located at the far end. This propagation delay is measured and compensated for during system startup to ensure complete synchronicity and simultaneousness of the received clocks by the devices. This action ensures that individual clocks for each device are contained to within 1μs of each other.

The local clock in each device ensures that the output signals from each device are outputted completely in synchronism. Slightest deviation from synchronism will lead to failure of coordinated action in multiple servo axes motion control. Error becomes more glaring because of short bus cycle.

The EtherCAT master must only ensure that the frame arrives at the individual devices ahead of application of distributed clocks at each device for position measurements for subsequent calculation of velocity.

18.6   REDUNDANCY

High redundancy in a network system leads to less downtime and consequently higher production will be ensured. If break in a cable occurs or a node fails or malfunctions, higher redundancy will ensure that the network does not fail.

A line topology in an EtherCAT network is converted into a ring topology by simply connecting the last node to an additional node in the master device. In such a case, a cable break or a slave malfunctioning will be detected by software incorporated in the master stack. For this to be realized, only a single Ethernet port and no special interface card is required. It is shown in Figure 18.3.

It takes a maximum of 15 μs to detect and restore disruption in the network in a line topology. Thus, at best one communication cycle will be involved which will not affect most machine operations significantly.

Master redundancy can be taken care of by simply keeping a hot standby.

18.7   DIAGNOSTICS

Networks which are rich in diagnostic features are more sought after by designers. Error localization is as important as error detection because localization of error pinpoints the source of error and thus the same can very quickly be resolved. This leads to higher machine availability.

Each node (slave) has an EtherCAT slave controller (ESC) in it which checks for error in the moving frame with the help of checksum. The node is allowed to advance to the next one in the network if no error is there, otherwise the error counter is incremented and subsequent nodes are informed accordingly. ESC detects this error and locates its source via the

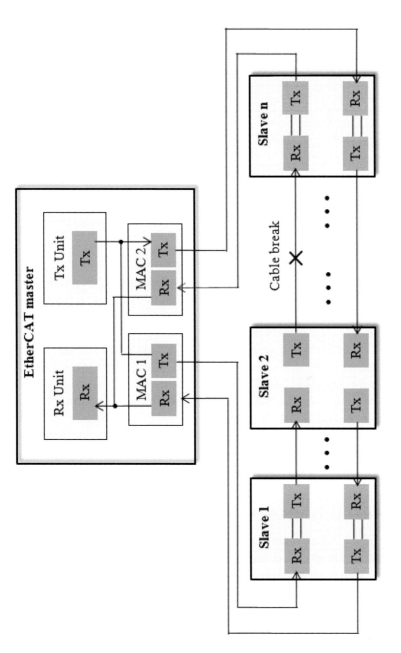

FIGURE 18.3 Line redundancy in an EtherCAT network.

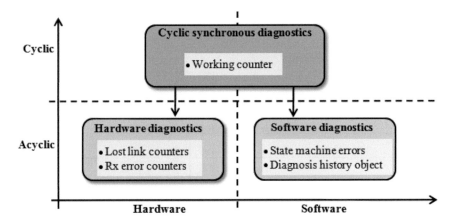

FIGURE 18.4   Hardware and software diagnostic functionalities in an EtherCAT system. (Courtesy: EtherCAT-The EtherNet Fieldbus, EtherCAT Technology Group. www.ethercat.org)

error counter. Subsequently, the frame is discarded. This is a major advantage compared to a fieldbus system where an error occurring is allowed to proceed, thereby making it extremely difficult to locate the source of error. This error localization, coupled with short cycle times of an EtherCAT frame, ensures error elimination in the quickest possible time.

Again, each datagram in an EtherCAT frame has a working counter whose content continues to be incremented as the frame passes through the nodes for data (either read or write purpose). If the counter has a value other than what it should, the frame is aborted and is not allowed to pass to the control application. The controller detects the error with the help of status and error information and actions taken accordingly. Figure 18.4 shows the diagnostic functionalities (both software and hardware) that is available in an EtherCAT system.

18.8   FIELDBUS MEMORY MANAGEMENT UNIT

Several EtherCAT devices can be addressed by a single Ethernet telegram with several EtherCAT commands. This improves data rate usage significantly. Deployment of fieldbus memory management unit (FMMU) in each slave further improves the data rate usage. Each FMMU is configured individually and the logical address space of the FMMUs is 4 GB. The FMMUs are integrated in the EtherCAT slave ASIC and enables individual address mapping of each device.

The memory management unit (MMU) present in processors can map memory pages. In addition to this, an FMMU supports bit-wise mapping. It allows the two bits of an input terminal to be inserted individually at any place within the logical memory address space.

If an EtherCAT command is sent that can read or write in a particular memory space, then instead of identifying the particular slave device, FMMU helps in inserting the 2-bit input terminal data at the right place within the data area. Other terminals, whose addresses are matched also, will insert their data at places earmarked for them. This simultaneous insertion of data in the frame that runs through the slaves saves a lot of time.

During the initial phase up process, the master can assemble process images and can exchange them subsequently via a single EtherCAT command. Additional mapping in the master is no longer required and process data can be assigned directly for control purposes.

18.9 SAFETY OVER ETHERCAT

In addition to fulfilling real-time data transfer, a communication system should be able to transfer safety critical data over the same medium. Such data are transferred by safety over EtherCAT mechanism, called FSoE or FailSafe over EtherCAT. Presence of FSoE in an EtherCAT system ensures (a) normal and safety critical data via the same medium, (b) same tools for normal and safe data, (c) both types of data are seamlessly integrated, and (d) diagnostic capabilities should be present for safety critical data. The safety technology was developed as per IEC 61508 and standardized in IEC 61784-3. The safety protocol has a safety integrity level up to SIL 3.

The safety over EtherCAT frame is shown in Figure 18.5. It is shown at the bottom and is encapsulated in the EtherCAT frame. Finally, the Ethernet frame encapsulates the EtherCAT frame.

EtherCAT provides the same bus or channel for transferring safety and non-safety data—called the black channel. Safety over EtherCAT frames—also known as EtherCAT container, contain safety critical data and also information to secure this data. This is shown in Figure 18.6. Both standard and safety applications are routed through the EtherCAT communication interface existing in a device.

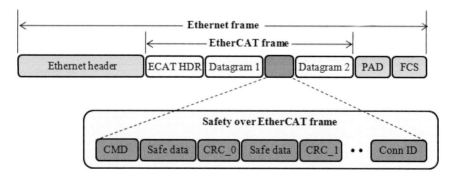

FIGURE 18.5 Safety over EtherCAT frame. (Courtesy: EtherCAT-The EtherNet Fieldbus, EtherCAT Technology Group. www.ethercat.org)

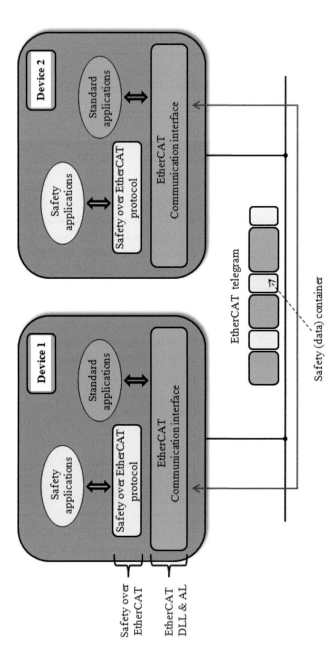

FIGURE 18.6 Black channel passes both safety and non-safety data. (Courtesy: EtherCAT-The EtherNet Fieldbus, EtherCAT Technology Group. www.ethercat.org)

18.10   ETHERCAT P

Protocol technology wise, there is no difference between standard EtherCAT and EtherCAT P (P for power). In the P mode, both communication data and power are sent via a single, standard four wire Ethernet cable.

EtherCAT P devices are powered by two electrically isolated, individually switchable 24 V power supplies. Several such devices can thus be cascaded and powered by a single cable. When EtherCAT P is used in a network, it would result in reduced cabling. It is particularly suitable for machine parts which are a bit isolated and placed at far off places. EtherCAT and EtherCAT P can be used together because technology wise both are identical.

18.11   ETHERCAT G

EtherCAT G, introduced in 2019, is an extension of standard EtherCAT protocol and operates at 1 Gb/s and 10Gb/s (EtherCAT G/G10). It is fully compatible with IEEE 802.3 and topologically identical with EtherCAT. The new version is suitable for applications in areas with huge process data like machine vision, complex motion operations, etc.

EtherCAT G introduces the concept of branch (segment), with each segment being controlled by individual EtherCAT branch controller (EBC). The segments operate in parallel at 100 Mb/s.

A network may have several EBCs and their configuration is managed by the master. The master must have a Gb/s port. Features associated with standard EtherCAT viz., diagnostics, network synchronization, and distributed clocks are passed on to the individual segments via the corresponding EBCs.

A frame is forwarded to a segment which is either priority or time controlled. Thus, separate segments can handle data processing in parallel. Parallel processing of individual network segments results in increased bandwidth and lesser propagation delays.

19 Sercos III

19.1 INTRODUCTION

An open and IEC compliant universal bus for Ethernet based real-time communication standard, SERCOS III (SErial Real time COmmunication System) is the third generation of the standard introduced in the year 2005, the first generation of which was introduced in the market in 1985. It is defined in IEC 61158, 61784, and 61800-7. It combines *on the fly* packet processing for delivering real-time Ethernet and standard transmission control protocol/Internet protocol (TCP/IP) communication to deliver low latency industrial Ethernet.

SERCOS was initially designed as a dedicated drive interface and since then universally applied in motion control interface. The user organization is SERCOS International e.V. which oversees and supports ongoing development and compliance and has more than 60 member companies.

It uses summation frame method and Open Software/Modified Ethernet. SERCOS operates without hubs and switches thereby reducing latency. Each station in the network has its own ASIC or FPGA. It processes data twice per cycle. Thus, devices can communicate with each other without going through the master.

It is a multi-protocol capable one which makes it possible to transmit various communication protocols simultaneously by using a common and uniform network architecture. As an example, SERCOS III devices can coexist with Ethernet I/P and TCP/IP devices without real-time capability or functionality of individual protocols being compromised.

It is capable of linking 511 devices and can operate in both slave-to-slave and controller-to-controller modes.

19.2 FEATURES

It is an open standard complying as mentioned above. Like PROFINET and Powerlink, it operates at 31.25 μs with jitter less than 1μs. SERCOS is compatible with Ethernet TSN. It has a flexible topology like line or ring—the latter one has redundancy without any additional hardware. Devices can be hot plugged into the network thereby increasing machine availability. Comprehensive diagnostic functions make troubleshooting very easy. Cross communication between master-to-master or slave-to-slave are possible. It can be seamlessly integrated with several fieldbuses.

Masters in SERCOS can be both hard and soft (open source license) types. When the hard master is in operation, CPU load is considerably reduced. Several Ethernet buses (like EthernetI/P, EtherCAT, PROFINET, Modbus/TCP, etc.) and Ethernet protocols (like TCP/IP, UDP, OPC UA, Webserver, etc.) can run in parallel with SERCOS real time (i.e., no tunneling needed). It provides fastest SIL 3 safety reaction time with black channel approach for CIP safety and AS-i safety.

19.3 TOPOLOGY

Every SERCOS III device has two Ethernet ports—one port connected to the just earlier one and the other connected with the one that just follows. They are connected via CAT5e Ethernet cable.

A SERCOS III network can be configured in either line or ring topology. Either of the topologies eliminates the need to have expensive network technology like switches. With line topology, a daisy-chain cabling is used and only one port of the master is connected to the first slave in the line. The slaves are connected in series. Data passes through the slaves one after another. The last slave in the line chain loops back to the master.

Ring topology supports redundancy. In this case, the primary port of the master is connected to the first slave and the last slave is connected back to the secondary port of the master. Both the ports of the master transmit the frames simultaneously. In case of cable break, the slave that detects the break, puts itself in the loopback mode. The break is detected by the master in the acknowledge telegram (AT) frame received by it. The master issues a ring heal command to the slaves to alter and restore the connection. Both the topologies are shown in Figure 19.1.

19.4 CONFIGURATION OF COMMUNICATION CYCLE

Predominantly used in motion control–based automation systems, SECOS III protocol combines the standard Ethernet with the need for real-time accuracy required in automation systems. It is based on a time slot mechanism with a cyclical transmission of telegrams based on master-slave (M/S) principle. Communication cycle times range from 31.25 µs. 62.5 µs, 125, µs, 250 µs with a maximum value of 65 ms.

The configuration of SERCOS communication cycle is shown in Figure 19.2. The cycle is divided into two time slots or channels—real-time channel (RTC) and unified communication channel (UCC). SERCOS defined real-time telegrams are transmitted through the RTC that is free of collision. Parallel to this, UCC is configured to pass Ethernet telegrams and IP based protocols (TCP/IP or UDP/IP). Communication cycle (bus cycle) and division of bandwidth are adjustable to accommodate RTC and UCC as per the need of specific application.

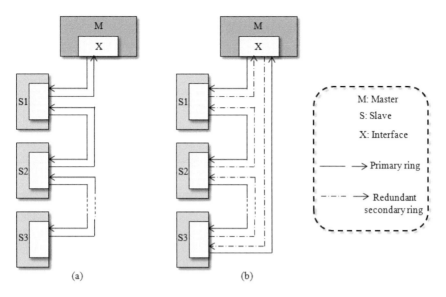

FIGURE 19.1   (a) Line structure (single ring). (b) Ring structure (double ring).

In the real-time channel, two telegrams are used: master data telegram (MDT) and acknowledge telegram (AT). Both telegrams are issued by the SERCOS master. MDT contains information provided by the master and read by the slaves. The AT is issued by the master, but actually is populated by the slaves with their response data. A slave uses the AT and fills the same in its pre-determined place, updating the checksum, before passing it to the next slave in the line. A communication cycle may consist of more than one MDT and AT (maximum four for each). MDT and AT are subdivided into three fields: hot plug field, service channel field, and real-time data field. The first one relates to slaves which are to be plugged in while the operation is on. The second one is the device channel for slave configuration and the last one is for device channels used by slaves to transmit real-time data.

SERCOS operates mainly in master-slave configuration for exchanging cyclic data between nodes. The master initiates all data transmissions during real-time channel. All data transmissions begin and end at the master. If any message is received by a node while the cyclic mode is active, the said message will be buffered and stored in the node and will be transmitted during UCC.

A SERCOS telegram with a clearly defined data structure is shown in Figure 19.3.

19.5   SYNCHRONIZATION

To realize hard real-time characteristics, SERCOS III uses a synchronization *mark* issued by the master at exact equidistant time intervals. This is used by all the nodes in the network to synchronize their operations.

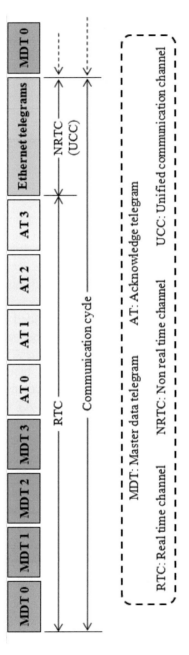

FIGURE 19.2 Configuration of SERCOS III communication cycle. (Courtest: SERCOS III Real Time Communication with Ethernet, Rexroth Bosch Group, DBR Automation. www.boschrexroth.com)

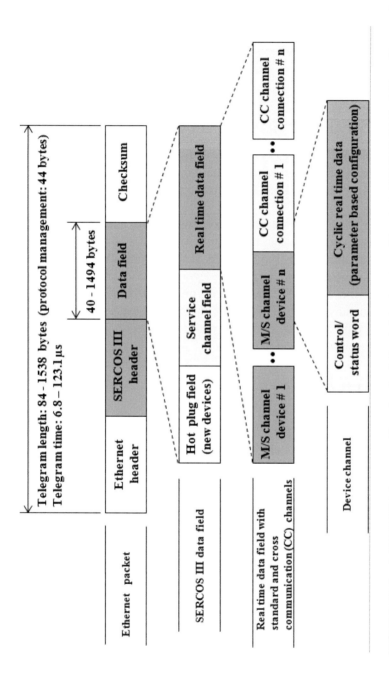

FIGURE 19.3 A SERCOS telegram with a clearly defined data structure. (Courtest: SERCOS III Real Time Communication with Ethernet, Rexroth Bosch Group, DBR Automation. www.boschrexroth.com)

Node-to-node transmission delays are measured during phase-up (initialization) and such delays are accommodated and compensated during normal operation. In SERCOS I and II versions, a separate master synchronization telegram (MST) was used. Unlike this, in SERCOS III, MST is not used separately but included in the first MDT of a communication cycle. The time interval between two successive MSTs is equal to the length of the communication cycle.

19.6   COMMUNICATION MECHANISM

Communication in SERCOS III has to fulfill determinism in real-time channel and is configured to pass Ethernet telegrams and IP-based protocols (TCP/IP or UDP/IP) in UCC. This combined mechanism is shown in Figure 19.4.

It shows three levels: hardware, communication, and application. The physical layer corresponds to hardware level and is a fast duplex Ethernet. At the top of this level is the SERCOS III communication controller that acts as the master. At the application level, communication is segregated into RC channel and UC channel. Real-time master-slave, service channel, and slave-to-slave communication are undertaken using RT channel. Non real–time communication like Ethernet-based telegrams and TCP/UDP-based protocols are passed in UCC.

19.7   HARD AND SOFT MASTER

A SERCOS master can be of two types: *hard* or *soft*. In a hard master, specific hardware is installed and is used in embedded applications (micro controller based motion control, drives, etc.). To keep the jitter within ns range, it is advisable that the overhead of managing the nodes should not be placed on the device processor.

In the soft master version, an operating system which is totally hardware platform-independent is used. Since the jitter in this case is dependent on the operating system of the master, a variable may be set in the network to limit the maximum value of jitter to be allowed.

19.8   ADDRESSING

Devices in a SERCOS III network need to comply with two separate addressing schemes: Ethernet's media access controller (MAC) addressing and SERCOS III addressing. Whether a device has an IP address or not depends on the need of the device to require such an address in its unified communication channel.

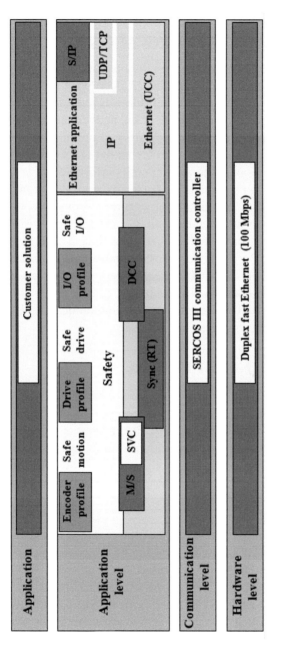

FIGURE 19.4 A combined communication mechanism to pass RT and UCC. (Courtesy: Plug and Play-The Sercos Automation Bus, www.sercos.org)

Each and every SERCOS III device requires a numeric address which will be used by other such devices in the network to exchange data. Such a numeric address may range from 1 to 511.

19.9 PEER-TO-PEER COMMUNICATION

Most real-time Ethernet standards follow a master-slave (M/S) way of communication to realize determinism. A node wishing to communicate with another node in the network will have to route its data/information via the master. This increases latency of the network and at the same time acts as an extra load on to the master. Direct communication between two slaves, by bypassing the master, may corrupt the synchronous operation of the system leading to an inefficient network performance.

SERCOS slaves can communicate between themselves directly via cross communication and at the same time ensures minimum communication dead times. It ensures unlimited real-time communication. A SERCOS III telegram passes each node twice in a cycle (this is true irrespective of topology—line or ring), a node has the opportunity to access data supplied by a subsequent node. Two peer-to-peer communication methods are specified here: controller–to-controller (C2C) for data/information exchange between two masters and cross communication (CC) for data exchange between two slaves.

19.10 COEXISTENCE OF ETHERNET
AND TCP/IP PROTOCOLS

Machine integration is increasingly becoming a tedious and tough job with proliferation of automation technology. Coexistence of emerging protocols with industrial Ethernet in a network architecture impacts real-time performance of the protocol in a majority of cases. But SERCOS III devices and Ethernet I/P devices can be connected via a single Ethernet cable without degrading performance and for this no additional hardware or tunneling is required.

Implementing a mix of these two technologies requires a SERCOS master and an Ethernet I/P scanner—they can be combined into what is called a dual stack master. When no redundancy is required, a line topology is used and shown in Figure 19.5. At the end of the line, when a SERCOS device identifies a SERCOS-unknown device on its second Ethernet port, it transmits non-SERCOS telegrams which are meant for other devices. In the reverse direction, the device transmits incoming telegrams to the dual stack master by using the first Ethernet port. For this, it uses the UC channel. Any Ethernet telegram coming during real-time operation will be buffered and retained for subsequent transmission.

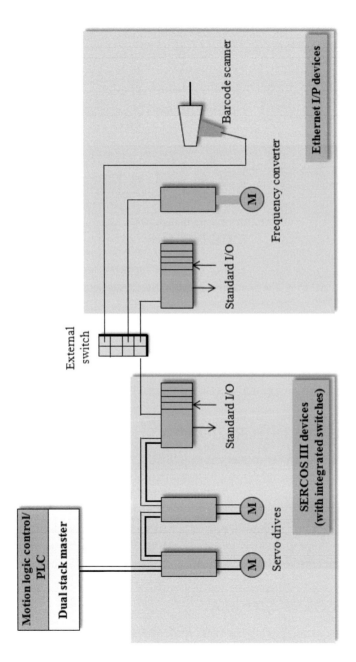

FIGURE 19.5 Combination of SERCOS III and Ethernet I/P devices using line topology. (Courtesy: Plug and Play-The Sercos Automation Bus, www.sercos.org)

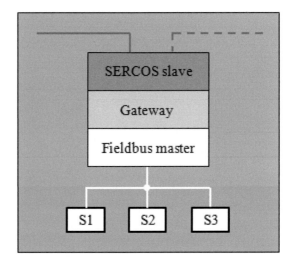

S: Fieldbus slave

FIGURE 19.6 Connection of a fieldbus system to a SERCOS network via gateway. (Courtesy: Plug and Play-The Sercos Automation Bus, www.sercos.org)

Redundant data transmission uses ring topology for which no free SERCOS port is available. For such cases, an IP switch is integrated into the ring or in a device. The switch either connects or disconnects the Ethernet I/P packets into the SERCOS ring. The Ethernet I/P devices can have either star or line topology between themselves.

19.11   FIELDBUS INTEGRATION

A fieldbus system can be connected to a SERCOS network via a gateway. This is shown in Figure 19.6, which shows that all the field devices (slaves) are connected to a fieldbus master. The fieldbus master is then connected to the SERCOS slave via the gateway.

A gateway can be an integral part of a SERCOS device or else can be a separate one. Communication gateways are available which can couple a fieldbus system (PROFIBUS, CAN, AS-i, etc.) to a SERCOS device

19.12   SERCOS VS. ETHERCAT

When it comes to comparing the two protocols, they have several things in common and certain significant differences do exist also.

Both use IEEE 802.3 as transmission medium. Both use summation frame method for efficient use of bandwidth. For faster operation both process data *on the fly* during real-time operations.

TABLE 19.1
SERCOS III vs. EtherCAT

Serial no.	SERCOS III	EtherCAT
1	Cross communication between slave-to-slave, master-to-slave and master-to-master is possible	Such cross communication is not possible here
2	No tunneling required. It is possible to connect any Ethernet device to a SERCOS network without additional hardware, resulting in reduced timeco	Tunneling required. Any Ethernet device requires tunneling before being connected to an EtherCAT network. This requires additional hardware.
3	Network wide synchronization is done directly from cyclical real-time protocol.	Time synchronization is based on distributed clocks. Such clocks require a time synchronization protocol which effectively reduces the bandwidth for user data.
4	It uses CIP safety protocol to transmit safety-related data.	It uses its own safety protocol—called FSoE (Fail Safe over EtherCAT).
5	It focuses on high production machines.	Used in a broad range of applications, mainly low-performance areas.
6	Membership not required to implement this technology.	Membership in EtherCAT technology group is free but mandatory in order to have access to technical specifications.

The differences that exist between them are shown in Table 19.1.

19.13   SERCOS OVER TSN

Standard Ethernet is not deterministic and hence not suitable for hard real-time communication. Using Ethernet TSN makes real-time critical message transmission much easier than in SERCOS III. Instead of using specific hardware, standard Ethernet components are used in Ethernet TSN. This is shown in Figure 19.7.

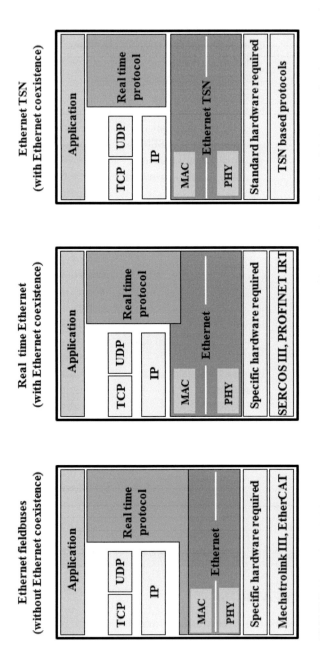

FIGURE 19.7 SERCOS over TSN. (Courtesy: Sercos over TSN, The Evolution of Industrial Communication moves forward. www.sercos.org)

Some of the advantages of using Ethernet TSN are: (a) Used in many sectors like automotive, automation etc. A range of manufacturers and a range of products is bringing down the cost substantially. (b) New technical developments with higher speed (c) IT and automation domain convergence is gathering pace.

20 Ethernet Powerlink

20.1 INTRODUCTION

Ethernet Powerlink (herein called EPL) is a real-time open protocol managed by Ethernet Powerlink Standardization Group (EPSG) and introduced by Austrian automation company B&R in 2001. It uses both mixed polling and time slice mechanism for deterministic data transmission. Data which is less time critical in nature like maintenance or configuration of a device, can also be sent in asynchronous mode. A gateway uses this mode to transmit non-Powerlink fieldbus data. The protocol enables integration of different networks—i.e., it is an industrial Ethernet solution designed to give users a single and integrated means for handling all communication tasks in today's automation systems. Hubs, and not switches, are used to limit EPL jitters.

It can be applied in process industry as well as machine and plant engineering. All the components in an industrial automation system viz., programmable logic controllers (PLCs), I/Os, human machine interfaces (HMIs), motion controllers, safety control, safety sensors can be integrated by Ethernet Powerlink.

Ethernet Powerlink is included in standards 61158-300, IEC-61158-600, and IEC 61558-500. There are two different versions of EPL. The first version is a proprietary of B&R while the second version was developed by EPSG.

Figure 20.1 shows the network diagram of EPL.

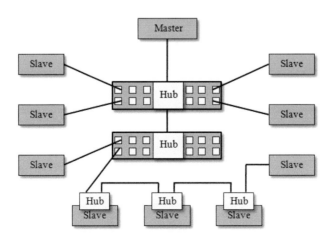

FIGURE 20.1 Powerlink network diagram.

20.2 FEATURES

Ethernet Powerlink is an open and standard compliant architecture whose development and standardization are maintained by EPSG. It is one of the fastest real-time Ethernet systems with cycle time around 200 μs and jitter below 1 μs obtained without any special hardware support. EPL devices can straightway talk to each other without a detour through PLC taking the help of either publish/subscribe or client/server relationship. This is particularly suitable in multi-axis drive applications. This cross-traffic communication between devices bypasses the master. It supports any network topology like star, tree, or daisy chain. EPL supports *on the fly*, i.e., hot plugging is possible. It implies that any node can be plugged in or disconnected without impairing network functionality or else no rebooting is needed. This facility gives Ethernet Powerlink an edge over its peers. Using mixed polling and time slice mechanism in EPL guarantees transfer of time critical data in very short and precise isochronous cycles with configurable timing and less time critical data in reserved asynchronous cycles.

20.3 POWERLINK ARCHITECTURE

The protocol architecture of Ethernet Powerlink is based on open systems interconnection (OSI) model and is shown in Figure 20.2. It supports both client/server and producer/consumer communication relationship. The protocol is based on IEEE 802.3 layers. Repeating hubs are used instead of switching hubs to reduce path delay and frame jitter.

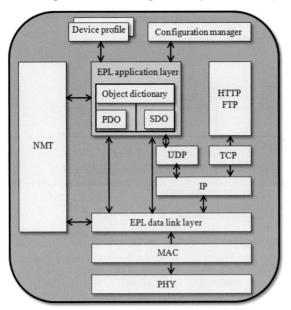

FIGURE 20.2 The Powerlink architecture.

20.4 OPERATION

Powerlink uses a mixed polling and time slot mechanism which ensures that a node is able to send data at a precise instant of time. EPL is a purely software-based solution that is IEEE 802.3 Ethernet compliant. The standard Ethernet data link layer is extended by an additional bus scheduling mechanism. The basic cycle (one clock cycle) consists of three phases: start phase, isochronous phase, and asynchronous phase.

A motion controller, an industrial PC, or a PLC acts as the managing node (MN) which behaves as the moderator in the network. It speaks to the other nodes (called controlled node (CN)) one after the other cyclically on a precise time schedule basis. MN defines the clock pulse for synchronization of all devices and manages data communication cycle in the network.

At the beginning of the clock cycle, MN sends out a start of cycle (SoC) frame to all CNs to synchronize the devices. The synchronization accuracy depends on (a) ability of MN to transmit SoC equidistantly, (b) ability of the CNs to time stamp SoC precisely, and (c) accumulated jitter of all hubs between MN and CNs. It is followed by isochronous phase where critical payload data is exchanged. The third phase is the asynchronous phase in which non-time critical data packets like parameterization data, etc. is sent.

In the isochronous phase, MN issues a poll request (Preq) to a node which responds with a poll response (Pres). This way data from all the nodes are accessed by MN in this phase which is an example of producer-consumer relationship. The time frame (Preq-n and Pres-n) is called time slot for the addressed node. The scheme of data transmission is shown in Figure 20.3. Frames above the time axis are governed by MN while below it are governed by CNs.

In the asynchronous phase, MN grants the right to a specific node to send ad-hoc data by initiating a start of asynchronous (SoA) frame. The addressed node answers with an Asnd. Standard IP based protocols and addressing are used during this phase to send device diagnostics and configuration data.

Time duration of the phases may vary but their total time must adhere to the basic cycle time which is monitored by MN. The duration of both isochronous and asynchronous phases can be configured.

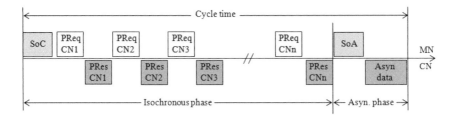

FIGURE 20.3 Data transmission in EPL by isochronous and asynchronous phases. (Courtesy: Powerlink Basics: System Overview, www.ethernet-powerlink.org)

The CSMA/CD media access control method, because of its non-deterministic Ethernet behavior, is not used in EPL.

20.5   ASYNCHRONOUS DATA

Non-time critical asynchronous data is sent in asynchronous frame. If a device has data which cannot be accommodated in a single asynchronous frame, then it is spread over several such successive frames. Routers/gateways are used to separate asynchronous data from the real-time domains of Ethernet Powerlink networks. Data flow in this phase is either way and hence devices can be accessed and configured with a real-time domain from outside. Various kinds of data can be sent during this phase: application data such as surveillance camera feeds, service data objects for diagnostic and configuration of devices. This phase is also used to integrate components or plant segments equipped with other types of fieldbus interfaces. For such a case, devices are tied to a Powerlink node via a gateway and data from such devices are sent in the asynchronous phase. Hot plugging (addition/deletion of devices during working of Ethernet Powerlink) is also done during this phase in which relevant data/information from such devices is sent to MN.

20.6   TOPOLOGY

EPL is an *any topology* network, i.e., a user can choose any topology for the network that would suit the purpose. Networks may have tree, star, ring, daisy chain, or a combination of any of these. Any topology change can be done on the fly without affecting network performance. Figure 20.4 shows a combination of the different topologies that can be employed in an EPL network.

20.7   MULTIPLEXING

Some of the controlled nodes need to be polled in every cycle for data. For example, in motion control applications, drives are to be provided with

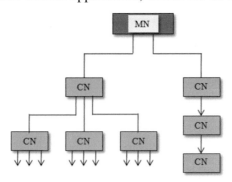

FIGURE 20.4   Different topologies employed in EPL. (Courtesy: Powerlink Basics: System Overview, www.ethernet-powerlink.org)

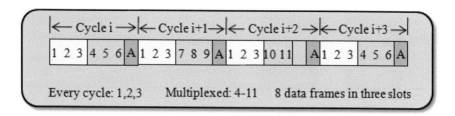

FIGURE 20.5 Multiplexing technique in isochronous phase. (Courtesy: Powerlink Basics: System Overview, www.ethernet-powerlink.org)

information every cycle. On the other hand, temperatures from sensors may be polled after every two or three clock cycles.

In multiplexing, controlled nodes with low priority data (like temperature) may share their time slots in the phase for better bandwidth utilization. Thus, an isochronous phase can be built which can distinguish between transfer slots dedicated to particular nodes which send their data continuously in every basic cycle, and slots shared by nodes to transfer their data one after the other in different clock cycles. Thus, less important but at the same time, time-critical data can be transferred in longer cycles than the basic cycle. The managing node has the discretion for allotting time slots for this multiplexing purpose. This is shown in Figure 20.5. In each cycle, time critical data from nodes 1, 2, 3 are accessed while less time critical data from nodes 4 to 11 are multiplexed for data accessing purpose.

20.8 REDUNDANCY

Ethernet Powerlink implements various types of redundancies: ring redundancy, partial ring redundancy, and master redundancy. Figure 20.6 (a) and (b) shows the connections for ring and partial ring redundancy techniques, respectively.

In ring redundancy, applications are connected to form a ring. The two ends of the ring are connected to the controller, i.e., PLC (which acts as an MN). The controller must have two redundant ports to support redundant

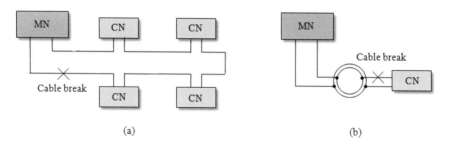

FIGURE 20.6 Redundancy: (a) Ring and (b) partial ring. (Courtesy: Powerlink Basics: System Overview, www.ethernet-powerlink.org)

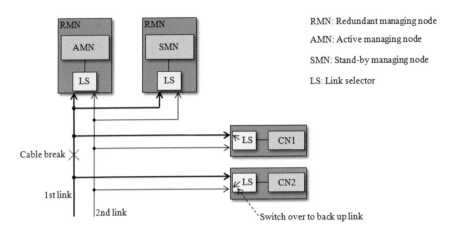

FIGURE 20.7 Master redundancy. (Courtesy: Powerlink Basics: System Overview, www.ethernet-powerlink.org)

operation. When the running data path fails, the system switches over to the redundant data path in one clock cycle.

Partial ring redundancy is applied to a specific part of a ring to safeguard some vital rotating applications. When the outer ring is affected, as shown in the figure, it switches over to the redundant path.

Master (managing node—PLC) redundancy is shown in Figure 20.7, which uses two or more MNs for its operation. It is used when high availability of data is a must. One MN remains active at any given time, the other MNs behave as CNs from the active MN's point of view. The standby MN continuously monitors the network and the CNs for any break of cable. If that happens, the standby MN immediately switches over and acts as MN from that instant, without any need to reboot the system. This redundancy scheme can be applied for a range of topologies.

20.9 SAFETY

Ethernet Powerlink is an open protocol with a high level of security so that a separately wired safety bus is not required. It has a very high visibility of all devices on the network allowing any device to be monitored, configured, diagnosed, or upgraded from any accessible network point.

EPL operates in a protected environment so that hackers would find it very difficult to penetrate through. EPL will connect itself with a non-deterministic Ethernet network via a gateway/router device, thus acting as a defensive barrier against malicious attacks. To achieve this, it employs a MAC address filtering and a built-in firewall. Barriers are placed in EPL so that only registered users can have access to real-time domains. It complies with various safety standards like IEEE, IEC, etc.

21 Profinet IRT

21.1 INTRODUCTION

PROFINET (PROcess FIeld NETwork) is an open standard in industrial Ethernet and finds widespread applications in areas like process automation, production automation, automotive, machine industries, and drive control. PROFINET is standardized in IEC 61158 and IEC 61784. The safety aspect of the devices are covered under IEC 62061/ISO 13849-1.

PROFINET is differentiated into different performance classes based on various timing requirements. The classes are: NRT (non-real time), RT (real time) and IRT (isochronous real time) and are based on producer/consumer principle and resorts to various protocols and services.

PROFINET IRT is mostly applied in motion control applications with cycle times below 1 ms. High priority payload data are sent via Ethernet protocol in Ethernet frames with VLAN prioritization. Cycle times can be optimized by using dynamic frame packing (DFP) principle and time multiplexed mode based on hardware synchronized switches.

The technology is around 25-year-old and developed by Siemens and other member companies of the PROFIBUS user organization, PNO.

21.2 FEATURES

Features associated with this technology are mainly: communication from management level to the field or device level, flexible topologies like star or line, flexible hard real-time communication that includes isochronous motion control, a dynamic frame packing principle which leads to flexible time cycles, supports a variety of transmission media like copper, wireless, fiber optic, etc., a seamless integration of all fieldbuses, safety incorporated for humans, devices, etc., prevents unauthorized entry of hackers into the system, a highly reliable diagnostic system, etc.

21.3 CONFORMANCE CLASSES

Three conformance classes, build upon one another, is defined in PROFINET. The classes are based on different requirements of automation systems. The classes are CC-A, CC-B, and CC-C.

CC-A is lowest in hierarchy with basic functions for PROFINET I/O and RT communication. IT services are also available in this class. In addition to functions provided under CC-A, CC-B provides functions like network diagnostics as well as network topology. Media redundancy protocol option is also available under this class. Applications include machine control with a deterministic but not isochronous communication. An extended version of CC-B is CC-B (PA) where system redundancy function as applied in process automation, is included.

CC-C includes all functions included under CC-A and CC-B. In addition, it provides highly deterministic real-time data communication using an isochronous phase. It has applications in machine control areas.

21.4   REAL-TIME COMMUNICATION— HARD AND SOFT REAL TIME

The PROFINET standard differentiates between three performance classes: PROFINET NRT (non-real time), PROFINET RT (real time), and PROFINET IRT (isochronous real time).

The first version is used in non-time critical applications that use transmission control protocol/Internet protocol (TCP/IP) or user datagram protocol/Internet protocol (UDP/IP) protocol for data transfer with cycle time around 100 ms. In the second one, I/O data is exchanged using Ethernet protocol but diagnostic and communication data is transferred using UDP/IP. Cycle time in this mode is around 10 ms. The last one, i.e., PROFINET IRT was developed for critical timing requirements in motion control applications with cycle time around 1 ms and jitter of 1 µs.

There is a difference between RT and IRT in PROFINET, although both versions refer to real time. In real-time industrial communication protocol, data exchange takes place within a specified time—typically of the order of less than 10 ms. Now, *real-time* systems can be hard real time or soft real time—it depends on how rigidly deadlines are enforced. Hard real-time systems are deterministic in nature—implying that the network guarantees that a message will be transmitted in a specified, bounded amount of time—it can neither be faster nor slower than that. In a hard real-time system, there is an absolute limit on response time. In a soft real-time system, on the other hand, occasional violation of the cycle time or deadline is accepted.

To realize real-time communication, PROFINET uses VLAN tag in the header. It sets the highest freely available priority level 6. It ensures that PROFINET telegrams are forwarded in preference via switches.

Among the three versions of PROFINET, PROFINET IRT corresponds to fastest data update rates. Now time needed for provisioning and processing of data is independent of type of communication undertaken for a system. Faster operation of data update is possible only through optimization of turnaround times in the stack. If faster stack operation can be done by bypassing some of the layers of open systems interconnection (OSI)

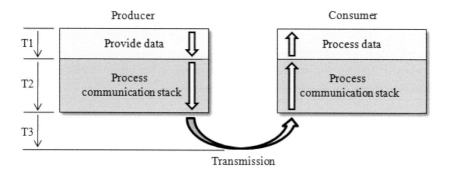

FIGURE 21.1   Different times needed in processing data. (Courtesy: PROFINET Real Time Communication. www.profinet.com)

model, it will lead to faster operation and turnaround time in the stack can be decreased. This is shown in Figure 21.1.

Times T1 and T3 remain constant whereas time T2 (when data passes through the stack) can be decreased by a combination of software and hardware.

Apart from PROFINET IRT, several other Ethernet-based industrial protocols like Ethernet Powerlink and SERCOS III achieve hard real-time communication by using the principle of isochronous data communication. Hard real-time systems operate on synchronized clocks to reduce cycle time and jitter and provide deterministic behavior.

21.5   REALIZATION OF FASTER OPERATION

PROFINET is an *open* protocol with short cycle time of 31.25 µs and jitter less than 1 µs for its IRT version. PROFINET V2.3 implements several functions that enhance its performance appreciably. Address information of a device is included in the frame header in case it is to be sent through the integrated switch. In such a case, the frame ID (FID) address information is integrated once in the corresponding switch. Thus, fast forwarding of the frame can take place reducing delay by several microseconds.

Another way to realize faster operation is *summation frame method*. In this, I/O data from several nodes (devices) are packed into one frame. Thus, only one FCS (frame check sequence) is required for several nodes leading to increased throughput. This is particularly advantageous for nodes having a few I/O bytes, because PROFINET uses 64 bytes in its frame like any Ethernet frame.

Enhanced performance can also be realized by employing a full duplex system in which both input and output data can be sent via the two pair cable. When a single summation frame is sent, received, analyzed, and checked down to the last node in the frame, DFP plays a crucial part. Since data already retrieved from the earlier nodes are not required at all by the latter ones, they are simply stripped off as the frame progresses through up to the last node in the frame. Thus, the frame becomes more and more shortened and arrival of data for the last node becomes much quicker. Thus, the cycle update time becomes much faster.

Another advantage of PROFINET V2.3 is its unlimited TCP/IP communication when IRT communication cycles are still on. The technique involved is accepting large TCP/IP frames in individual nodes and fragmenting them. The individual fragmented parts are sent in consecutive cycles. At the receiver, they are assembled together at the application layer to get back the unaltered TCP/IP frame. This makes possible to realize bus cycles of 31.25 µs duration in a shared I/O and TCP/IP communication. The integration of the fragmented parts is done in the switch module and thus does not require any additional special device.

21.6 WORKING OF IRT

Working of IRT is based on time slice mechanism. Traffic on the IRT network consists of both IRT and RT—it is assumed that IRT traffic is allocated 25 percent of the total network bandwidth and the rest is assigned for RT traffic, as shown in Figure 21.2.

IRT traffic passes for one-time slice fulfilling stringent time schedules. During this time, any non-IRT traffic would be buffered. Once the IRT traffic is over, the buffered traffic would be allowed to pass with the help of switches and regular Ethernet communication will take place. The reserved IRT traffic is scalable—i.e., it is just large enough to accommodate IRT communication.

Highly precise time slices (for both RT and IRT) can be achieved by switches. This is possible if the network has the following:

1. An extremely accurate master clock which will synchronize all the connected devices across the network to create the time slices of exact duration.

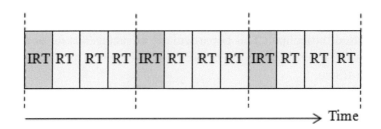

FIGURE 21.2 Data traffic on IRT network: Time slice mechanism.

2. The switches must have some additional circuitry which would buffer and hold any Ethernet traffic that may be received during the execution of IRT phase.

The master clock is based on IEEE 1588v2 (IEEE 1588-2008) which defines the precision time protocol (PTP). This is put in a protocol wrapper called precision transparent clock protocol (PTCP). Delays inherent in network switches and cabling can be calculated by PTCP. The high precision real-time clock across the network along with having a very precise delay calculation allows the switches in the network to enter and exit the IRT time slice exactly at the required time. The clock master uses synchronization frames to synchronize all local clock generators inherent in such devices. The devices are connected directly to one another, without going through any non-synchronized devices.

Figure 21.3 shows how data in PROFINET protocol passes through the different layers of OSI. There are three variants of data: standard data, RT data, and IRT/TSN data. RT and IRT communication bypass session, transport, and network layers, leading to faster turnaround times in RT and IRT.

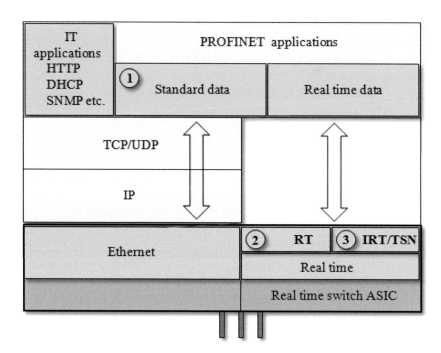

FIGURE 21.3 PROFINET variants follow different ways to pass through OSI layers. (Courtesy: PROFINET Real Time Communication. www.profinet.com)

21.7   TIME SENSITIVE NETWORKING (TSN)

It is a series of new standards whose aim is to improve determinism in *standard* Ethernet networks. Industrial Ethernet is an open standard and Siemens is working with PROFINET at the field or device level and OPC UA at the control level. Siemens is presently working with Ethernet-based standard time sensitive networking for reserved bandwidth with main emphasis put on quality of service, low transmission latency, and parallel transmission of different protocols in real-time domains of industrial networks.

At present, TSN offers cycle time of 31.25 μs and a jitter of 1 μs, but PROFINET IRT already offers such performance levels. PROFIBUS and PROFINET International (PI) have already adopted TSN in the latest PROFINET standard.

TSN improves upon the existing time synchronization clock (IEEE 1588) by deploying another new standard 802.1AS-2019 which has some additional features. TSN networks adopting such new standards will generate an error if synchronization goes out of expected bound. This facility is absent in IEEE 1588. Another feature associated with the new standard is that it has priority of scheduling. With TSN, there is a now a standardized layer 2 in the OSI model which is upward compatible to the previous Ethernet and hard real-time capability.

TSN embraces an extensive network configuration with both centralized and decentralized mode of operations possible. Interoperability between these two modes is being developed presently. With introduction of TSN, layers 1, 2, and 3 of OSI model will be unified into one with higher levels of scalability and performance becoming a reality.

21.8   USING IRT

Every equipment in PROFINET IRT (also called PROFINET conformance class C)—starting from controllers to devices and switches must be compliant with PROFINET conformance class C. The minimum data update rate in this class is 250 μs and a jitter of less than 1 μs. The data update rate can be brought down to 31.25 μs using proper hardware. For update rates less than 250 μs, TCP/IP communication is fragmented and transmitted in smaller packets.

One needs to properly configure PROFINET IRT before using it. This requires specifying the number of time slices (bandwidth) needed for IRT operation, and in addition cycle times for the devices. Again network topology must be clearly specified which would enable the IRT devices to optimize the IRT transmission schedule within the IRT time slice. Configuring a class C network is a bit more lengthy and complex than a class B network.

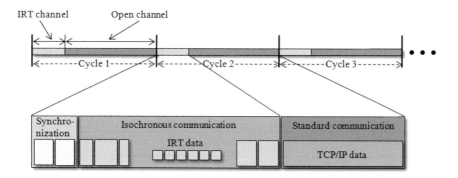

FIGURE 21.4   Different schedules in PROFINET IRT communication. (Courtesy: PROFINET Real Time Communication. www.profinet.com)

Some controllers, devices and switches support *Isochrone Mode Application* although it is not part of IRT. This mode is used to lock the application's execution loop to the IRT update interval. This prevents over or under sampling, ensuring entry to or exit from IRT time phase to be perfect.

Figure 21.4 shows the schedule that is followed in PROFINET IRT communication. A particular cycle begins with synchronization of all the devices in the network. This is done by the highly accurate master clock. As per the need of the connected network, IRT data is sent in reserved bandwidth (time slices), followed by RT communication. Diagnostic data, etc. are sent under the head *standard communication.*

22 Intrinsically Safe Fieldbus Systems

22.1 INTRODUCTION

Fieldbus standard IEC 61158-2 defines the maximum dimensions for a fieldbus segment and its proper operation in a safe area. These operations and calculations are based for a network based on FOUNDATION fieldbus H1 or PROFIBUS PA (MBP). Some additional constraints are imposed for proper operation of the network segment in the hazardous area for explosion protection as defined in IEC 60079.

22.2 HAZARDOUS AREA

In chemical industries in general and petrochemical industries in particular, which deal with oil and gases, a high degree of caution is always maintained because of the presence of flammable gases and oils. Heat, electrical arcing are the main reasons for fire/explosion in these hazardous areas. Thus, presence of oxidizer (air/oxygen), flammable gases, and ignition energy (in the form of thermal or electrical energy) contribute to sparks taking place in industries. This is commonly known as *ignition triangle* and shown in Figure 22.1. Safety of plant and personnel is of prime importance in such industries. Intrinsic safety, explosion-proof enclosures, purging are some of the measures undertaken to overcome fire hazards.

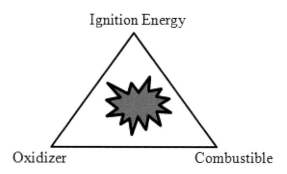

FIGURE 22.1   The ignition triangle.

22.3   HAZARDOUS AREA CLASSIFICATION

Classification is based on identification and quantity of flammable materials, their characteristics, and chances of their coming in contact with arcing, etc. The basis includes types of flammable gases/vapors, liquids, combustible dust, and fire-prone fibers.

Depending on the classification, equipment employed in hazardous areas match and conform to the safety standards. The standards are classified in National Fire Protection Agency (NFPA) standard 70—the National Electrical Code. Normally two classifications are followed: the conventional Division Classification System and Zone Classification System.

22.3.1   DIVISION CLASSIFICATION SYSTEM

In this, a hazardous area is identified with a class, division, and group. *Class* deals with whether the flammable material is a gas or dust or fiber, while *division* deals with the probability of presence of flammable material by defining how it is used. Finally *group* deals with the physical properties of the flammable material.

There are three classes: I, II, and III. Class I deals with presence of flammable gases, class II with combustible gases, and class III with ignitable gases.

Classes are further subdivided into groups. The groups assign categories of chemical and physical properties to the hazardous materials. These are found in NFPA and NFPA 325.

Class I is subdivided into four groups: A, B, C, and D. Group A consists of acetylene while groups B, C, and D are concerned with hydrogen, ethylene, and propane gas, respectively. Class II is subdivided into three groups: E, F, and G. Information about class II group are formed in NFPA 499. Class III does not have any subgroup.

Division is concerned with probability of flammable materials present and its use in the hazardous area. It is subdivided into Division 1 and Division 2.

22.3.2   ZONE CLASSIFICATION SYSTEM

It classifies a hazardous area into three zones: Zone 0, Zone 1, and Zone 2. They were introduced in 1999 by the National Electrical Code. Zone 0 area refers to presence of flammable mixtures of gaseous vapors continuously or for long periods of time. In zone 1, flammable mixtures are expected to occur under normal circumstances while in Zone 2, flammable mixture is unlikely under normal circumstances and would quickly disperse if they do occur.

22.4   EXPLOSION PROTECTION TYPES

Figure 22.2 shows both division-wise and zone-wise hazardous area classifications.

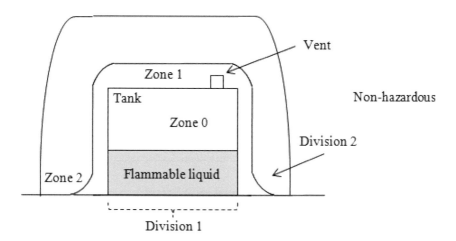

FIGURE 22.2 Hazardous-area classifications. (Courtesy: Fieldbus Wiring Guide, 4th Edition, Austin, TX, Doc. No. 501-123, Rev.: E.0. www.relcominc.com/pdf/501-123FieldbusWiringGuide.pdf)

22.5 INTRINSIC SAFETY IN FIELDBUD SYSTEMS

Topologically, a fieldbus network and an intrinsically safe fieldbus network do not differ. However, the latter network includes lesser number of devices per wire and a safety barrier is placed instead of regular power supply impedance. Field instruments for intrinsically safe systems are designed such that they do not inject power into the bus.

A terminator is placed on either side of a fieldbus network. The network contains a capacitor which must be certified to be intrinsically safe. Also the terminators must be totally passive which would draw no current.

Again for each intrinsically safe segment, only one power supply source is employed—i.e., no redundancy is allowed. The barriers are placed in safe areas to reduce cost. Sometimes exigencies force the barriers to be mounted in hazardous areas, in which case they must be housed in flame proof enclosures with flame proof seals.

The barriers employed can be a zener barrier or a galvanic barrier. Fieldbus devices operate between 9 V and 32 V. Thus, to limit power, lesser voltage should be chosen for powering field devices in order to limit power consumption. It is a good investment if field devices are chosen which consume least power. In such cases, lesser number of barriers would be needed.

Figure 22.3 shows a safety barrier that separates a safe area from a hazardous area.

Intrinsically safe fieldbus barriers have been around since 1960s and different barriers have been introduced over a period of time as shown in Figure 22.4. These are: Entity Concept, FISCO, HPTC, and DART.

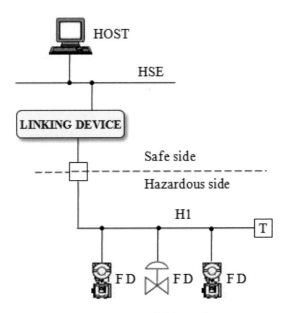

FIGURE 22.3 A safety barrier separates a safe area from a hazardous area. (Courtesy: B. G. Liptak. Instrument Engineers' Handbook, Process Software Digital Network, 3rd Edition, CRC Press, Boca Raton, FL, p.8611, 2002)

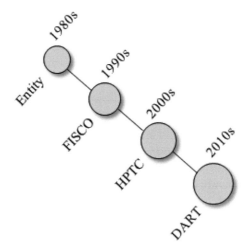

FIGURE 22.4 Development of different intrinsically safe fieldbus systems. (Courtesy: A. Beck and A. Hennecke. Intrinsically Safe Fieldbus in Hazardous Areas, Technical White Paper, EDM TDOCT-1548_ENG. Pepperl+Fuchs GmBH, Mannheim, Germany, p. 8, 2008. files.pepperl-fuchs.com/selector_files/navi/productInfo/doct/tdoct1548a_eng.pdf)

22.6 ENTITY MODEL

It is defined in IEC 60079-11 and NEC 515 and is a method of validating an intrinsically safe installation by employing intrinsically safe parameters. The parameters to be considered are voltage, current power, capacitance, and inductance. The cable capacitance and inductance for the hazardous side of the segment are considered lumped and must be considered along with the capacitance and inductance of all the devices connected to the segment. Several devices are normally multi dropped off a single barrier and hence the entity parameters of all the devices must match those of the barrier.

A linear output characteristic for the barrier is used in entity concept, shown in Figure 22.5. The output power for Exia IIC is roughly 1.2 W at 11 V DC, the current available is a maximum 60 mA. This thus limits the maximum number of devices to a few for each barrier otherwise available voltage at the devices would become less than the minimum recommended one. Application of small voltage (11 V DC) limits the total cable run, since a small voltage drop along the cables would bring down the voltage available at the devices to a value less than the recommended one. A barrier allows a maximum L/R ratio which should be compared with that of the cable. Again, cables with shielding have higher capacitance, which lowers the total cable run compared to cables without shielding.

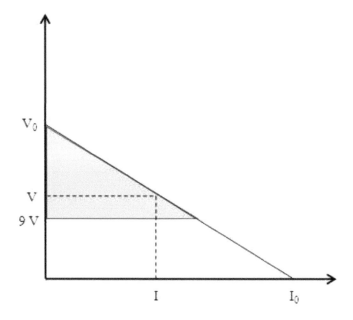

FIGURE 22.5 The linear output characteristic. (Courtesy: J. Berge. Fieldbuses for Process Control: Engineering, Operation and Maintenance, ISA, p. 103, USA, 2004.)

Fault disconnection electronics (FDE) automatically disconnects the power supply to the barrier in case of malfunctioning/short circuit of any device.

The low power of 1.2 W allows a maximum of 2 to 3 devices per barrier for IIC gas connection. This practically limits adopting the entity concept in hazardous areas. Again this method is complex and needs time-consuming calculation efforts to validate an installation.

22.7 FISCO MODEL

Fieldbus Intrinsically Safe Concept (FISCO) was developed by PTB (Physikaliseh-Technische-Bundesanstalt) and has trapezoidal characteristics, shown in Figure 22.6. It provides an output power for Exia IIC. Thus, more devices can be connected per segment compared to entity model. FISCO offers the following advantages compared to entity model: (a) increased available power, (b) simplification in calculations, and (c) installation parameters are standardized.

FISCO recommends a single power supply per fieldbus segment and other devices present in the segment are only power drains and no power feedback in the cable is allowed.

Some FISCO models offer 1.2 W output power and effective for devices with lower power ratings. Such FISCO equipment are often referred to as *small FISCO* or *fisco*.

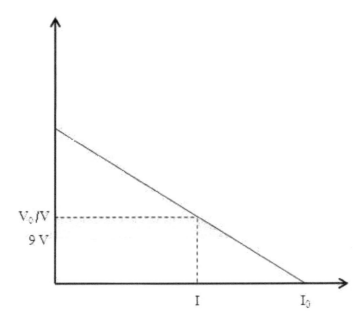

FIGURE 22.6 The trapezoidal output characteristic. (Courtesy: J. Berge. Fieldbuses for Process Control: Engineering, Operation and Maintenance, ISA, p. 106, USA, 2004.)

Barriers, devices, and cables in FISCO model must be FISCO compliant. FISCO restricts the use of a single power supply per network, thus power supply redundancy is not possible.

FISCO does not require any calculations and thus offers one of the easiest methods for validation of explosion protection. It shifts the onus of electric design from the planner and operator of process plants to the equipment manufacturers. For practical situations involving short cable lengths, 4–8 devices can be connected per segment depending on the gas group. On the flip side, FISCO requires expensive power supplies and low mean time between failure (MTBF) due to complicated circuit designs.

The ratings for FISCO power supplies are around 250 mA (Group C, D, IIB) and 110 mA (Group A, B, IIC).

22.8 REDUNDANT FISCO MODEL

The difference between FISCO and redundant FISCO lies in power supplies. Two arbitration modules, shown in Figure 22.7, ensure that only one power supply out of the two intrinsically safe power supplies is active at any given instant of time. The two modules always monitor the output voltage of both the power supplies.

When output voltage of one falls below a specified value, the module ensures the switch over of the power supply to the other one. During the switch over, the bus loses power and the segment voltage drops. As a rule of thumb, a maximum of 100 ms is allowed to avoid field devices resetting

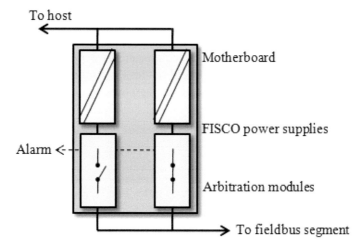

FIGURE 22.7 Redundant FISCO power supply arbitration principle. (Courtesy: A. Beck and A. Hennecke. Intrinsically Safe Fieldbus in Hazardous Areas, Technical White Paper, EDM TDOCT-1548_ENG. Pepperl+Fuchs GmBH, Mannheim, Germany, p. 8, 2008. files.pepperl-fuchs.com/selector_files/navi/productInfo/doct/tdoct1548a_eng.pdf)

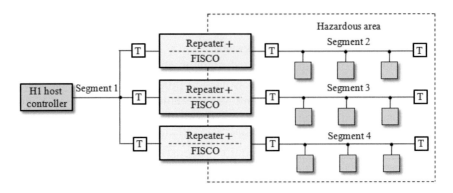

FIGURE 22.8 Multi drop FISCO power supply configuration. (Courtesy: Fieldbus Wiring Guide, 4th Edition, Austin, TX, Doc. No. 501-123, Rev.: E.0. www.relcominc.com/pdf/501-123FieldbusWiringGuide.pdf)

due to power loss. However, due to redundancy, communication telegrams are lost due to power drop.

Redundant FISCO, although ensuring power supply to the segment, has drawbacks like higher capital cost, more cabinet space, probability of communication errors occurring during redundancy power transfer, and the two arbitration modules requiring their own power putting constraints on cable length and number of devices per segment.

22.9 MULTI DROP FISCO MODEL

A multi drop FISCO power supply configuration is shown in Figure 22.8. In this, several FISCO power supplies are multi dropped to different segments—with each power supply connected to a particular segment. In situations where the number of field devices is limited due to intrinsic safety considerations, connecting several segments results in utilization of full capacity of DCS' interface card. Thus, considerable cost savings are effected in DCS hardware.

22.10 HPTC MODEL

High Power Trunk Concept (HPTC) was introduced and developed in 2002 by Pepperl+Fuchs. It removed the limitations with regard to segment length and number of devices. In HPTC, more power is available to the segment which requires that the trunk cable be protected in the form of armoring or putting the same in a duct—i.e., mechanical methods of explosion protection are effected in the hazardous area. Field devices are connected to the trunk cable via intrinsically safe (IS) barriers to reduce power to intrinsically safe levels and shown in Figure 22.9.

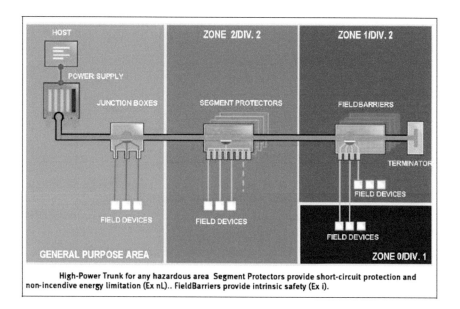

High-Power Trunk for any hazardous area Segment Protectors provide short-circuit protection and non-incendive energy limitation (Ex nL).. FieldBarriers provide intrinsic safety (Ex i).

FIGURE 22.9 Block diagram of HPTC for hazardous area. (Courtesy: A. Beck and A. Hennecke. Intrinsically Safe Fieldbus in Hazardous Areas, Technical White Paper, EDM TDOCT-1548_ENG. Pepperl+Fuchs GmBH, Mannheim, Germany, p. 8, 2008. files.pepperl-fuchs.com/selector_files/navi/productInfo/doct/tdoct1548a_eng.pdf)

Compared to other models, standard power supplies are employed in HPTC. These power supplies are easily available and less costly. The power supplies are employed in redundant configuration. The two power supplies are in parallel and share the load evenly. In case of one power supply failing, the other immediately takes over. Also, since under normal circumstances both the power supplies share the load, their life expectancy obviously increase. The attributes of HPTC are: (a) highest possible overall cable length, (b) redundancy in power supply, (c) easy validation of intrinsic safety requiring no calculation, (d) Entity and FISCO compliant devices can be mixed in a segment, (e) field devices can be serviced without 'hot work' permit, and (f) standard power supplies can be employed.

22.11 DART MODEL

Dynamic Arc Recognition and Termination (DART) attacks the power availability in intrinsic safety applications with a completely new approach compared to its predecessors. DART ensures considerably higher available power during normal conditions, while conforming to intrinsically safe energy limitations via rapid disconnection during emergency situations.

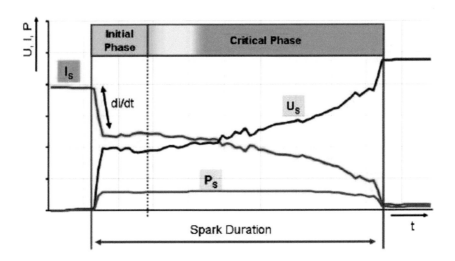

FIGURE 22.10   Electrical behavior of spark. (Courtesy: A. Beck and A. Hennecke. Intrinsically Safe Fieldbus in Hazardous Areas, Technical White Paper, EDM TDOCT-1548_ENG. Pepperl+Fuchs GmBH, Mannheim, Germany, p. 8, 2008. files.pepperl-fuchs.com/selector_files/navi/productInfo/doct/tdoct1548a_eng.pdf)

DART realizes long cable runs and many devices per segment without the need for protected installation, as in the case of HPTC.

Sparks in wiring in electrical systems can cause explosions of hazardous gases. Sparks may occur during *make* or *break* of electrical circuits. Enough energy in the spark may cause the existing gas to attain ignition temperature, leading to explosions. Typical behavior of spark is shown in Figure 22.10 which shows that it remains non-incendive during initial phase, but reaches critical phase within several microseconds and become incendive.

All the attributes of HPTC are present in DART. Along with that, the trunk is intrinsically safe which allows maintenance along the trunk without hot work permit.

DART does not limit power during normal operation, as is the concept prevalent in the other ones. On the other hand, it detects a fault condition by the characteristic rate of change of current and disconnects power before it becomes incendive, leading to catastrophic consequences.

A DART compliant power supply feeds the system with full 8 to 50 W, compared to approximately 2 W as in Entity concept. When a fault occurs, DART detects the resulting rate of change of current di/dt (that occurs in the initial phase of the spark) and switches off power supply to the system in approximately 5 ms, before the critical phase or incendive condition arises. This is shown in Figure 22.11. Thus, power to the system is reduced to a safe level, spark is robbed of its energy, and the system does not become incendive.

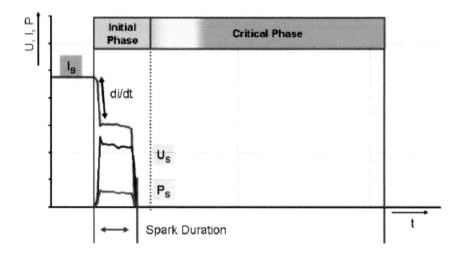

FIGURE 22.11 The highly characteristic di/dt associated with a spark. (Courtesy: A. Beck and A. Hennecke. Intrinsically Safe Fieldbus in Hazardous Areas, Technical White Paper, EDM TDOCT-1548_ENG. Pepperl+Fuchs GmBH, Mannheim, Germany, p. 8, 2008. files.pepperl-fuchs.com/selector_files/navi/productInfo/doct/tdoct1548a_eng.pdf)

The rate of change of current *di/dt* is highly deterministic—thus a fault condition is easily detected and the system is saved from catching fire.

DART fieldbus is very simple and straightforward, requiring no additional training for field personnel. During the planning phase, a potential user will only have to keep in mind the maximum trunk and spur lengths permissible. DART service personnel need not have to distinguish between *black* and *blue* cables—as is the common practice with field barriers.

DART is very simple to install and maintain, entailing great savings both in terms of capital and operating costs. DART fieldbus is certified by PTB as per IEC 60079-11 with an IEC-Ex certificate.

DART supports 32 devices per segment which is the maximum permissible one. This thus reduces excessive infrastructure and reduces capital cost. Thus, lesser number of junction boxes, power supplies, barriers, etc. is required to implement a DART fieldbus system.

22.12 PERFORMANCE SUMMARY

Table 22.1 shows the basic merits and demerits of the four mostly used intrinsically safe fieldbus systems, viz., Entity, Redundant FISCO, HPTC, and DART. It can be concluded from the table that both HPTC and DART have almost identical performance levels, but live working on the trunk is possible only in DART.

TABLE 22.1

General Performance Summary of Different IS Modules

	Entity	FISCO (Redundant)	High power trunk	DART
Available power	-	0	+	+
Validation of explosion protection	-	+	+	+
Power supply redundancy	-	- (+)	+	+
Long term physical layer diagnostics	-	-	+	+
Segment design mix	-	-	+	+
Cabinet space requirement	-	- (- -)	+	+
Power supply initial cost	-	-	0	0
Trunk live working	+	+	-	+

Source: A. Beck and A. Hennecke. Intrinsically Safe Fieldbus in Hazardous Areas, Technical White Paper, EDM TDOCT-1548_ENG. Pepperl+Fuchs GmBH, Mannheim, Germany, p. 8, 2008. files.pepperl-fuchs.com/selector_files/navi/productInfo/doct/tdoct1548a_eng.pdf

22.13   CONCLUSION

Several measures like intrinsic safety, explosion proof enclosures, etc. are employed for safety of plant and personnel in hazardous industries. Uninterrupted operation of fieldbus based instrumentation systems is a must to keep downtime to a minimum and enhance profitability. Over the years, different types of intrinsically safe fieldbus systems have been introduced, the latest one being the DART. HPTC and DART offer almost comparable benefits to users of fieldbus systems. DART offers around 8 watts per segment which is about 4 times more than that provided by FISCO. Since live working on the trunk is possible only with DART, coupled with its ability to provide higher levels of available power, it is increasingly becoming the most sought after technology where intrinsic safety is a must for proper fieldbus system operation.

23 Wiring, Installation, and Commissioning

23.1 INTRODUCTION

In a process industry, field devices are installed, wired together, and commissioned properly before they can be put into operation. Electrical installation wise, FOUNDATION Fieldbus and Profibus PA are identical since they conform to IEC 61158-2. Again HART, FOUNDATION Fieldbus, and PROFIBUS PA use existing wires for their proper operation. HART devices are discussed separately since their installations differ on several counts.

It is recommended to consult manufacturer's data sheet before installation and commissioning of any fieldbus system. Fire hazards employing fieldbus is a major area of concern in certain process industries and has already been discussed in Chapter 22 on *Intrinsic Safety in Fieldbus Systems*.

23.2 HART WIRING

Highway addressable remote transducer (HART) is an open access process control network protocol. The HART protocol uses frequency shift keying (FSK) technique to superimpose digital signal on the conventional 4-20 mA current loop and hence sometimes termed as a hybrid communication. A 0.5 mA sine wave of two frequencies, viz., 1200 Hz and 2200 Hz representing binary 1 and 0, respectively, are superposed on the 4-20 mA DC signal. Thus, it adds no DC component as the average value of a sine wave over a complete cycle is zero. Data is transferred at a rate of 1200 bps.

HART uses different topologies for connecting field devices to the controllers. It can be point-to-point, multiplexed, or multi-drop types, shown in (a), (b), or (c) of Figure 23.1. Irrespective of topology, an FSK modem is used at the output of a field device and also at the input to the computer which may act as a host or a primary master.

HART protocol uses master-slave communication. In this, a master initiates communication and the slave merely responds to it. The masters can be a PC-based network controller, hand-held terminal, distributed control system (DCS), or a programmable logic controller (PLC). There can be two masters in a HART network at the same time. A remote terminal unit (RTU) or a multiplexer can act as an interface for the primary master.

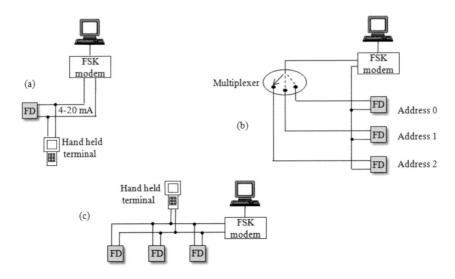

FIGURE 23.1 HART network topology: (a) Point-to-point, (b) employing multiplexers, and (c) multi-drop mode.

A hand-held terminal is a secondary master and is connected only when working in the field. Sensors, actuators, transmitters act as slave devices.

A multiplexer allows several fieldbus devices to be connected to the host one after another, with all the field devices having the same address. Data and status information from field devices are put into the host for processing. Several multiplexers can be cascaded so that many devices can be accessed by the host—all the connected devices having the same address. Here, multiplexer acts as a gateway.

23.2.1 SURGE PROTECTION

Lightning strikes and occasional starting and stopping of electrical motors are of serious concern in industries and the equipment must be protected from such surges. Surge voltages are of the order of tens of KVs and lasts only a fraction of a second. Surge suppressors are used to protect devices which include employing low capacitance in the device. The capacitance acts an open circuit to the segments and the spurs and hence does not affect communication in a fieldbus system. Such protection methods are identical for different fieldbus types.

Transmitters with potential to surge strikes should be placed in separate segments with no final control elements. In case a surge strikes and the transmitter is affected, the consequent data loss will not affect the system much than a process upset resulting from both the transmitter and the final control element becoming offline due to a surge strike.

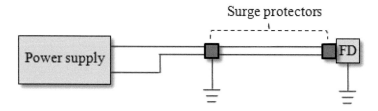

FIGURE 23.2 Surge protectors installed at either end of a network. (Courtesy: J. Berge. Fieldbuses for Process Control: Engineering, Operation and Maintenance, ISA, p. 72, USA, 2004.)

Surge protectors are installed at field devices of a segment and at the host. Also the number of field devices per segment should be kept to a low value so that in the event of a strike, the consequent process upset can be kept to a minimum. It should also be borne in mind that installation of surge suppressors must not attenuate the original fieldbus signal. Surge protectors are activated only when surge current exceeds a certain threshold. The extra current flows into the earth—thus a solid earth connection is a must. A surge protection method is shown in the Figure 23.2 which shows employing the protectors at either end of the system.

Again current limiting couplers are used in conjunction with surge suppressors. In the event of a surge strike, the suppressor would cause failure of the coupler, resulting in data loss and control of the affected segment.

23.2.2 DEVICE COMMISSIONING

Two addresses identify a HART device: a unique hardware address and a polling address. The unique hardware address consists of the manufacturer's code, the device type code and a unique identifier. The combination of these three codes ensure that no two hardware addresses are same. This eliminates any chances of conflict. The hardware address can never be changed.

The polling address is set to zero for point-to-point mode, while it is from 1 to 15 for multi-drop mode. For multi-dropping case, each field device is configured separately with individual polling addresses before being connected to the network. This address assignment is done for each device by connecting them point-to-point with the help of a hand-held terminal. Communication is not possible in case duplication of address is inadvertently done even for two devices.

23.3 BUILDING A FIELDBUS NETWORK

Fieldbus uses a pair of wires to power the field devices and also carry the process signals to the local controller. Different topologies are employed to carry the bidirectional (from devices to host and vice versa) digital signals

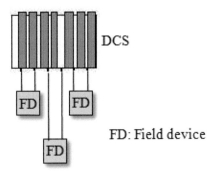

FIGURE 23.3   A conventional DCS system. (Courtesy: Fieldbus Wiring Guide, 4th Edition, Austin, TX, Doc. No. 501-123, Rev.: E.0. www.relcominc.com/pdf/ 501-123FieldbusWiringGuide.pdf)

in a fieldbus system which constitutes a local area network (LAN). Amongst the different fieldbuses available for different uses, FOUNDATION Fieldbus and Profibus PA are the mostly used ones for process control purposes. They share identical networking and powering schemes, although differing significantly in communication protocol strategy.

For a conventional DCS, process parameters like temperature, pressure etc. from sensors and control signals to actuators, valves etc. are carried via a shielded twisted pair cable in the form of 4-20 mA current signals via I/O cards. Figure 23.3 shows several process sensors connected to a DCS system via I/O cards.

The wiring diagram for a fieldbus system is shown in Figure 23.4 in which Figure 23.4(a) shows a two wire connection while Figure 23.4(b) shows the same in a simplified manner.

Figure 23.5 shows a single fieldbus device connected to a fieldbus network. Two terminators, one at each end of the wire pair, are connected.

23.3.1   MULTI FIELDBUS DEVICES

Figures 23.6 and 23.7 show a number of field devices added to the network with *star* connection for the former and *chained* network for the latter. In both the figures, only two terminators are attached, identical to Figure 23.5.

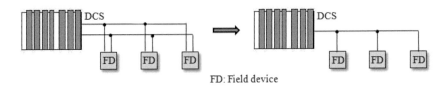

FIGURE 23.4   A fieldbus wiring diagram.

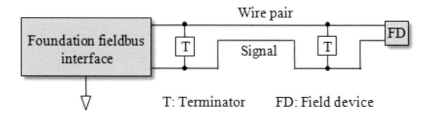

FIGURE 23.5   A fieldbus network with a single device.

In Figure 23.6, three field devices are joined at the junction with the terminator placed at the junction itself. It is tacitly assumed that all the three devices are about the same distance from the junction. Had one device been placed at a larger distance from the junction, then the terminator would have been shifted to the end of that device. In Figure 23.7, as devices are continued to be added to the chain, the terminator is shifted and connected across the last field device.

23.3.2   EXPANDING THE NETWORK

It is necessary to extend the reach of a network when the plant to be controlled is physically spread over a considerable area. In such cases network traffic regulators (also called network devices) such as hubs, repeaters, switches, bridges, routers, gateways are used to realize the above.

For traffic to flow in an orderly manner in a network, the system follows some protocols comprising hardware and software. NICs, hubs, switches, bridges, routers help in efficient communication over the network. These devices are called network hardware devices and help in regulating the speed on the network. If the hub operates at 10 Mbps, then the other network devices would operate at the same speed. Also these devices manage the flow of traffic by opening, closing or directing the traffic along a

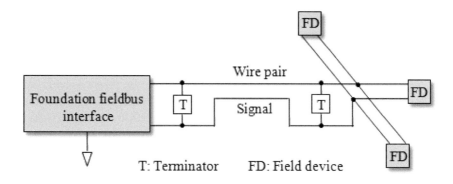

FIGURE 23.6   A star connected fieldbus system.

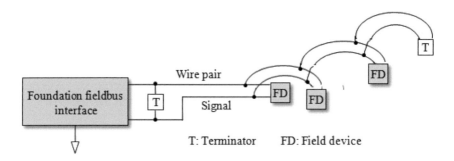

FIGURE 23.7   A chained fieldbus network.

specific route. For example, a router ensures the most efficient way for the traffic to reach its destination. A third job of these devices is to protect some sensitive devices.

Network communication takes place in a layered fashion. The open systems interconnection (OSI) model has seven layers with each layer performing a specific networking function. Protocols govern these functions and manage end-to-end communication between devices. As user data is passed from the upper layers down below, each layer adds a header (and sometimes a trailer also) to data. These headers contain protocol information and are termed as protocol data units (PDUs). The process of adding headers to the layers is known as encapsulation.

23.3.2.1   Network Interface Cards (NICs)

These are also known by adapters or network interfaces or simply cards. It is a device that connects a server, a client computer, printer, or any other component to the network. It is connected to the network via a small receptacle, called a port. For wired systems, the network cable is inserted into this port. For wireless networks, the port includes a transmitter/receiver that sends/receives the radio signals. NIC services include connecting the computer physically to the network and secondly it converts information on the computer to and from electrical signals for the network. For example, information on the computer is converted into electrical signal of appropriate shape and speed compatible to the network to which it is connected.

NICs connected to the network must conform to the physical and data link level protocols for the electrical signals to be compatible and information exchange to take place effectively and successfully. If the Ethernet network runs at 10 Mbps, then the servers, printers attached to the network must have 10 Mbps compatible NICs. Each NIC connected to the network must have a unique media access control (MAC) address which helps routing information within the LAN.

23.3.2.2   Hubs

A hub is a layer 1 device and physically connects network devices together. Hubs do not have any intelligence and thus a hub forwards every frame out of every port, only excluding the port from which the frame is originated. Hubs cannot differentiate between frame types and would thus forward unicast, multicasts, or broadcasts out of every port, except the originating one. Hubs are unable to process layer 2 or 3 information and thus cannot make any decision based on hardware or logical addressing.

Ethernet hubs operate in half duplex mode. Ethernet employs carrier sense multiple access with collision detect (CSMA/CD) to access a media before sending a frame and transmits the same only if the link is idle. If two devices send their frames at the same time, a collision will occur and both the frames are rejected after informing the devices which sent the frames. When more than one station want to send a frame, a back-off algorithm sees to it that a single station wins the bus arbitration. Thus a particular station gains access of the network and transmits a frame without any kind of error. All ports belonging to the same hub belong to the same collision domain. Hubs also belong to the same broadcast domain.

All devices (clients, servers, peripherals, etc.) connected to a hub share the bandwidth of the network and belong to the same collision domain. As more devices are added to a hub, probability of collision increases, thereby degrading performance. Probability of collision increases with increase in network traffic.

Hubs are divided into two types: passive and active. In passive hubs, signal is forwarded as it is, while in active hubs, signal is amplified before sending. Thus, an active hub behaves as a repeater. Hubs can also be cascaded to increase the network capacity.

23.3.2.3   Repeaters

In Foundation Fieldbus, segment length is limited to 1900 m due to attenuation along the length of the wires. Repeaters are used to increase the length of a network. A repeater refreshes the timing of a signal and boosts its amplitude to the original level at its input. A fieldbus controller chip does the above jobs. A repeater is bidirectional in nature and galvanically isolated. A maximum of four repeaters can be employed on a single network, thereby increasing the length of the network to 9500 m—thus having five segments. Schematically it is shown in Figure 23.8.

It makes economic sense to employ repeaters to connect intrinsically safe segments thereby forming a larger network with a corresponding increase in device count, rather than having a single safety barrier per each host communication port. Some typical field instruments consume a considerable amount of current, thereby limiting the number of such devices to

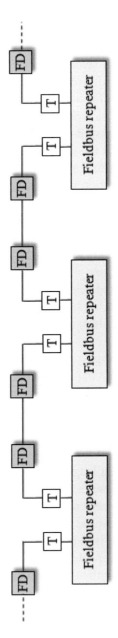

FIGURE 23.8 Repeaters in a fieldbus network.

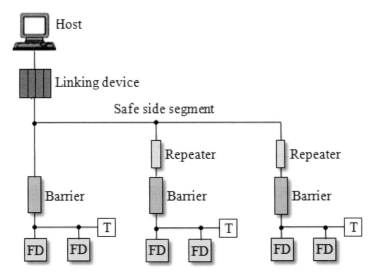

FIGURE 23.9   A large network formed with repeaters and barriers. (Courtesy: B. G. Liptak. Instrument Engineers' Handbook, Process Software Digital Network, 3rd Edition, CRC Press, Boca Raton, FL, p. 166, 2002)

two or three per segment. In a typical connection, a safe area segment can be connected to the host via linker on one side, while several multi-dropped barriers with repeaters attached, can be connected at the other side. A hazardous area segment is connected via a barrier, as shown in Figure 23.9.

A segment has two terminators, placed at each end of the segment. Repeaters with built-in terminators are available such that external terminators are dispensed with. Care should be exerted when using repeaters—they should be disabled or enabled as per requirements. Type of repeater to be used depends on the network. Thus, a repeater conforming to IEC 61158-2 must be used for proper system operation.

Repeaters can be active bus powered or non-bus powered devices. Device count can be increased to 240 by using repeaters. Repeaters operate at the physical layer of OSI and extend the physical length of a LAN. A repeater is a two port node—when it receives a frame from any port, it regenerates the original bit pattern and simply forward it to the other port. A repeater is a regenerator of data and not an amplifier.

23.3.2.4   Switches

A switch is a layer 2 (data link layer) device and makes intelligent forwarding decisions based on header information of this layer. It derives this intelligent decision based on special hardware circuits, known as application specific integrated circuits (ASICs). Only then, the frame is forwarded to the appropriate destination port and not to all ports.

A switch builds a hardware address table which contains a hardware address for the host devices and a port address that each device is attached to.

Ethernet switches build MAC address tables following a dynamic learning process methodology. When a switch is powered on first time, it behaves very much like a hub. It then floods every frame, including unicasts, out every port sans the originating one. The switch will start building the MAC address table by examining the source MAC address of each frame. As the MAC address table becomes more and more populated with time, the flooding of frames would continue to decrease, and the switch becomes more and more efficient.

A drawback associated with switches, like hubs, is its susceptibility to switching loops. This generates a destructive broadcast storm. It can be avoided by a spanning tree protocol. Thus, a switch has the following advantages which a hub does not have: loop avoidance, intelligent frame forwarding, and hardware address learning.

A switch operates in full duplex mode. Each individual port on a switch belongs to its own collision domain and hence a switch creates more collision domains resulting in fewer collisions. A layer 2 switch belongs to one broadcast domain while a layer 3 device separates broadcast domains. A switch cannot differentiate one network from another but can discriminate one host from another. Thus, switches are not suitable for large networks, which a router, belonging to layer 3 is capable of.

A switch has the capability to momentarily connect the sending and receiving devices so that they can use the entire bandwidth of the network without interference. Using a switch does not result in any benefit when the network traffic is poor. This is because a switch examines the information inside each signal on the network to determine the address of the sender and the receiver before sending the frame. This results in slow processing of signals and leads to latency.

A switch forwards frames using any one of the three methods: The store and forward, the cut through (real time), and the fragment free (modified cut-through).

23.3.2.5 Bridges

A bridge is an active bus powered or non-bus powered device. It connects two or more local area networks or two or more segments of the same network. It enables data from one network to be transferred to another having different speeds or physical layers—like wires, optical fiber, etc. Thus, a bridge enables different networks to talk to each other. Devices residing on different segments but on the same network must have different node addresses. But two devices on different networks may have same node address. It is the responsibility of the bridges to ensure that devices are not

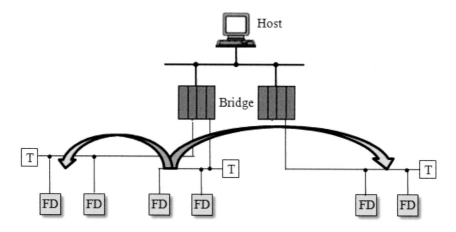

FIGURE 23.10 Different networks can talk to each other via bridges. (Courtesy: J. Berge. Fieldbuses for Process Control: Engineering, Operation and Maintenance, ISA, Research Triangle Park, NC, p. 99, 2004.)

mixed up. Bridges are normally built into a linking device or an interface card. Figure 23.10 shows bridges talking to each other.

Bridges are simple and efficient traffic regulators. Bridges, like switches, filter network traffic before forwarding them. Thus, overall traffic congestion is avoided. Again, like switches, bridges learn the MAC address of the connected servers, peripherals, clients, and associate each address with a bridge port. Bridges are layer 2 devices.

23.3.2.6 Routers

Routers are layer 3 devices and forward a packet from one network to another, based on the contents of the network layer header. This forwarding is based on the destination network and not on the destination host. The forwarding path is based on a routing table that contains the following: routing metrics and administrative distance, the destination network and the subnet mask, and the next hop the router to follow to reach the destination network.

Since each individual interface on a router belongs to its own collision domain, routers create more collision domains resulting in fewer collisions. Also routers separate broadcast domains. A router never forwards broadcasts and usually will not forward multicasts.

A router normally copies packets in its buffers and checks the router lookup table before forwarding the same. It thus results in latency. Now a days improved router versions use hardware to offset the latency effect.

Routing, in a way, is similar to switch but a router is not a switch. A switch connects two computers to form a LAN while routers connect two LANs, two WANs, or a LAN and its own ISP network.

Routers filter network traffic so that any unauthorized entry is denied. Routers ensure the best route that a packet is to take to reach the destination by avoiding congested routes. A router ensures that (a) traffic meant for a local network does not flood the Internet and (b) a traffic existing on the Internet which is not meant for the local network stays on the Internet and does not intrude the local network.

When a router receives a packet, it notices its hardware MAC address. If this address is not on a local segment, then this address is stripped by the router and it looks for the software IP address in its routing table On the basis of this comparison, the packet is then sent to the router that contains that address.

23.3.2.7 Gateways

A gateway performs protocol conversion and therefore networks with different protocols can be connected together via a gateway. Gateways are protocol specific and thus proper gateways must be employed for any message to be transported effectively from one network to another one. For example, data from a FOUNDATION Fieldbus HSE network can be transported to a PROFIBUS network provided a gateway is employed which converts a protocol from the former system to the latter one.

A gateway is manually configured by mapping the device address, data type or any other kind of information in the foreign protocol to the desired information format for every parameter that is to be sent to the foreign network.

23.3.2.8 Routers vs. Gateways

A router operates at the network layer (layer 3) while a gateway operates at the transport layer (layer 4). Both are used to regulate traffic on two or more separate networks. Routers regulate traffic between similar networks (protocol dependent) while a gateway regulates traffic between dissimilar networks (protocol independent.). A local network using Windows 2000 can be connected to the internet via router because both use TCP/IP as their primary protocols. While a device which uses a Windows NT network can be connected to the NetWare network via a gateway.

While routers are configured by their routing tables, gateways are configured by internal and external networks. Gateways act as a network point which acts as an entry point to another network. Router is a device which guides the data packet to its proper destination that arrives at the gateway. A gateway provides the entry/exit point to/from a network like the door of a house.

23.4 POWERING THE FIELDBUS DEVICES

Fieldbus networks carry both DC power supply to power the field devices and communication signal at 31.25 kHz from the devices. The lines carry both DC superimposed with AC. A normal power supply having both voltage

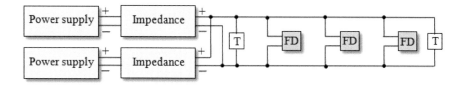

FIGURE 23.11   A conventional power supply along with an impedance module. (Courtesy: J. Berge. Fieldbuses for Process Control: Engineering, Operation and Maintenance, ISA, Research Triangle Park, NC, p. 100, 2004.)

regulation and noise reduction circuitry won't serve the purpose here, as such power supplies, having almost zero output impedance would short the 31.25 kHz communication signal. The power supply must have output impedance conforming to IEC 61158-2 such that the communication signal passes through the two lines without attenuation and preventing a short circuit. On the other hand, this output impedance of the power supply would offer almost zero resistance to DC and hence the current drawn by the devices would cause little drop on the network. Thus, either an IEC 61158-2 compliant power supply should be used or an impedance module compliant to IEC 61158-2 should be inserted between the power supply and the network.

Figure 23.11 shows a normal power supply and a IEC 61158-2 compliant impedance module connected to the network. A terminator is attached to the impedance module and another at the far end of the cluster of devices.

Redundant power supplies should be employed to ensure higher system availability. In such cases, it should be ensured that in case of failure in primary power supply, there is bumpless switchover to the secondary power supply. The power supplies must be galvanically isolated, short-circuit protected and must have a failure indication.

The 250 Ω resistor employed in HART plays an identical part as the impedance module here. This resistance prevents the power supply from short circuiting the HART signal operating at 1200–2200 Hz. At the same time, this resistance acts as a shunt which converts the modulating current into a voltage that can be picked up by the receiving devices.

Bus powered field devices belonging to FOUNDATION Fieldbus and PROFIBUS PA operate at voltages in between 9 and 32 V DC. Typically, power supply voltage output is around 24 V DC. Most field devices consume around 12 mA while some devices consume twice as much. Fieldbus power supplies provide a load of around 300 mA.

23.5   SHIELDING

Fieldbus instruments are placed at considerable distances apart even for a mid-sized plant. Because of the constant operation of heavy duty motors and

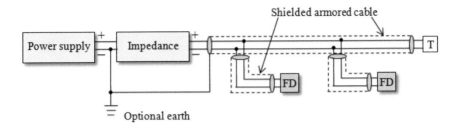

FIGURE 23.12   Shielding of spurs and trunks.

other electrical appliances, noise of varying degree is always present along the communication paths. The network cable paths therefore be shielded from such stray electromagnetic interferences. The cables may also be placed in a metal conduit which also provides mechanical protection. Spurs and trunks should be shielded and connected to the negative of the power supply, shown in Figure 23.12, and not at the negative of the impedance module.

Signal conductor lines should not be grounded because it may result in loss of communication. Usually shields are connected to ground at one point. If multiple grounding is required because of existence of potential differences along the ground path, capacitive shielding is preferred. In this case, grounding is made through a capacitor. Thus, only high frequencies in the noise get grounded.

23.6   CABLES

Fieldbus signal, as it travels down via spurs and trunks, is attenuated until it becomes too weak to be recognized properly. Thus, it puts a restriction on the maximum cable run which depends on the type of cable employed. Typical cable lengths and their types are specified in IEC 61158-2 to maximize backward compatibility of the existing cable infrastructure.

Multi twisted-pair cables are employed in large projects which helps reduce cable installation cost. Such cables support multi-fieldbus trunks. This is shown in Figure 23.13, where the multi-pair cable runs from the control room to a strategically placed junction box.

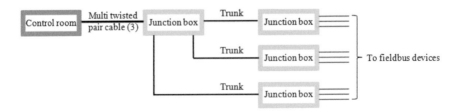

FIGURE 23.13   Use of multi-twisted pair cables.

Cable length restrictions are imposed by device power consumption considerations as also intrinsic safety. In such cases, the cable lengths become even shorter.

Fieldbus was originally designed such that existing cables in a plant can be utilized. In fieldbus systems, bus powered devices are mostly used. The field devices draw current and there is voltage drop along the wire. This limits the wire length that can be employed in any particular case, because the devices cannot operate below 9 V voltage level. For maximum cable length and the number of devices, supply voltage should be as high as possible, device current consumption as low as possible and wire cross section as high as possible to reduce resistance along the wires, thereby entailing lower voltage drops.

23.7   NUMBER OF SPURS AND DEVICES PER SEGMENT

The short cable length that branches out from the main trunk and connects to a device is called a spur. Cable lengths lesser than 1 m in length is termed a splice. Spur length is independent of the type (i.e., quality of cable) of cable but depends on the number of devices connected to it and the topology employed. Total spur length is determined by the number of spurs and the number of devices per spur.

Spur length can vary between 1 m and 120 m, although normal spur length should be limited to within 10 m. Spur length beyond 10 m, if needed, should be kept outside high-risk areas. Special attention should be placed for tree topology, because in this case each branch is a spur. If certain spur lengths are appreciably high, the junction box should judiciously be placed so that any spur length is limited to within 120 m. For intrinsically safe installations, the spur length is limited to 30 m.

For tree topology, each device connection from the junction box is a spur. In such a case, the junction box is placed in such a manner that the recommended maximum spur length is adhered to.

23.8   POLARITY

In powered fieldbus networks, alternating voltage from the field devices is superimposed on the DC voltage, while if the network is not powered, only alternating voltage from the field devices exists. Manchester coding used by the system ensures that polarity change will take place in each bit period. The fieldbus signal is thus polarized and all the field devices must be so connected that they all see the signal in correct polarity.

There are non-polarized network powered field devices which can be connected in either polarity across the network. These devices sense which terminal is positive and which is negative and have automatic polarity detection and correction capability. Such a device can receive a message of either polarity.

If a network consists of different types of devices, signal polarity must be taken into account with all positive terminals connected to each other, likewise the negative terminals also. Wires are color coded to make such connections easy. It is advisable to develop a network with polarity considered so that non-polarized devices can be blindly connected. It helps to add polarized devices later when expansion of plant necessitates addition of devices.

Signal polarity and power polarity must be the same for bus powered devices, while non-polar bus powered devices can accept power and signal of either polarity.

23.9   SEGMENT VOLTAGE

A fieldbus power supply is specially designed for the network to operate properly. Bus powered field devices draw their operating power in the same way as a two wire analog device. The farthest device in the segment of the network must be provided with a minimum voltage of 9 V. While designing a network with many field devices having trunks and associated spurs, the following must be taken into consideration:

1. Supply voltage
2. Current consumption of each field device
3. Resistance of every part of the network
4. Location of the power supply

DC circuit analysis is used to determine the voltage at each field device. Since a network consists of so many branches at different points in the network, calculation of field device voltages often becomes cumbersome. This becomes more so when some field devices are added to the network later at different places as per future requirements. To obviate such difficulty, fieldbus vendors have designed software that makes such calculations just a click away. Again increased resistance due to higher temperature must be kept in mind.

23.10   LINKING DEVICE

A linking device connects field-mounted devices to the host by means of higher speed host level network. It connects the FOUNDATION Fieldbus H1 to higher speed HSE and PROFIBUS PA network to PROFIBUS DP. They have buffers within to take care of speed differences and act as interface.

A linking device for FOUNDATION Fieldbus is shown in Figure 23.14.

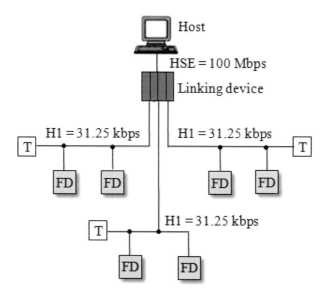

FIGURE 23.14 A linking device connects field devices to host. (Courtesy: J. Berge. Fieldbuses for Process Control: Engineering, Operation and Maintenance, ISA, Research Triangle Park, NC, p. 97, 2004.)

Redundant field devices are installed in parallel and two data paths are maintained to the host where very high data availability is a must. In the event of a linking device failing, the other takes over. The situation is shown in Figure 23.15.

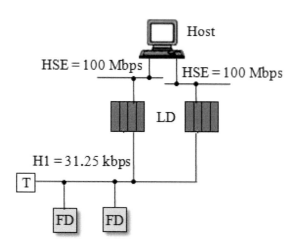

FIGURE 23.15 Redundant linking devices enhance data availability. (Courtesy: J. Berge. Fieldbuses for Process Control: Engineering, Operation and Maintenance, ISA, Research Triangle Park, NC, p. 98, 2004.)

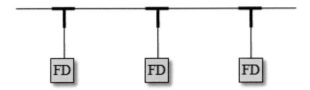

FIGURE 23.16 A simple **T** connection.

23.11 DEVICE COUPLER

A device coupler connects the spur cables to their corresponding trunks. FF-846 is the standard for device couplers for FOUNDATION Fieldbus. Compliance to this standard ensures better device coupler performance that includes input impedance, voltage drop, reaction to short circuit, etc.

There are two types of device couplers, viz., non-isolated and isolated. Device couplers route the field transmitter outputs to the bus and to the host controllers and then routing the commands back to the control devices. Device couplers provide short circuit protection against the whole segment failing if either one single device fails or the spur cable is shorted.

The simplest fieldbus device connections to the segments are via the spurs and its simplest version is T. But its problem is that if a device shorts itself, it would bring down the entire segment along with it. Such a connection is shown in Figure 23.16.

Figure 23.17 shows the schematic diagram of a device coupler. It shows two device couplers—one an 8-spur and another a 4-spur, connected via a short jumper cable. One end of the terminator is grounded for surge suppression. Thus, 12 field devices can be connected by such a device coupler. It is housed in a junction box.

Device couplers have current limiters built into them. This is shown in Figure 23.18. This prevents a *shorted* spur from short circuiting its corresponding segment.

Second-generation device couplers that have of late been introduced, have features like auto termination, short circuit protection, and visual circuit checking. They have quick up time and low maintenance costs. Such device couplers help accelerate device commissioning and routine maintenance by providing easy access points for hand-held communicators for ease of trouble shooting.

23.12 COMMUNICATION SIGNALS

As fieldbus signal travels along the cable, it is both attenuated and distorted. It is attenuated as it travels through the cable and also at the spur cable where it branches off to the trunk cable. The latter attenuation is caused mainly by cable capacitance. The transmitted and received fieldbus signals are shown in Figure 23.19.

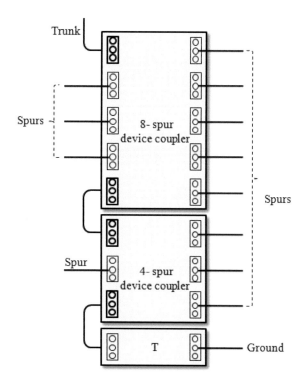

FIGURE 23.17 A schematic diagram of device coupler. (Courtesy: Fieldbus Wiring Guide, 4th Edition, Austin, TX, Doc. No. 501-123, Rev.: E.0. www.relcominc.com/pdf/501-123FieldbusWiringGuide.pdf)

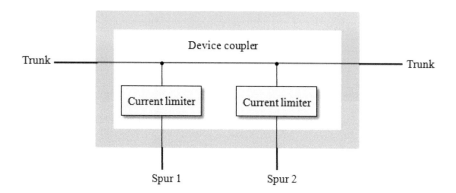

FIGURE 23.18 A device coupler with current limiters. (Courtesy: Fieldbus Wiring Guide, 4th Edition, Austin, TX, Doc. No. 501-123, Rev.: E.0. www.relcominc.com/pdf/501-123FieldbusWiringGuide.pdf)

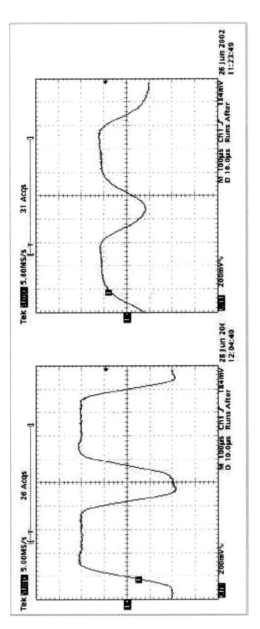

FIGURE 23.19 Transmitted and received fieldbus signals. (Courtesy: Fieldbus Wiring Guide, 4th Edition, Austin, TX, Doc. No. 501-123, Rev.: E.0. www.relcominc.com/pdf/501-123FieldbusWiringGuide.pdf)

Fieldbus signal quality is dependent on characteristics of cables, reflections at spurs, etc. Received signals are degraded if trunk length or total spur lengths are not adhered to as per recommendations.

23.13   DEVICE COMMISSINING

Device address assignment for FOUNDATION Fieldbus is done automatically by the linking device or some other interface while the same for PROFIBUS is configured manually before the device is installed on the network. Once a device is installed, it will show up in the live list, like all other devices, in the host application.

Device configurations can be started with the help of the engineering tool even before the arrival of devices at the site, and after the detailed engineering is over. Once the devices have been received at the site, the devices can be connected to the network and the device configuration files downloaded immediately. Such an approach cuts down the startup appreciably.

There are two possibilities when a device does not show up in the live list when it is connected to the network. The first is that the device is not connected properly and needs troubleshooting. The second possibility is that the device may appear in the live list of another network.

Prior to field device commissioning, several checks must be undertaken for proper working of the system. These are: (a) continuity of cables, proper grounding, and insulation testing, (b) field device couplers, (c) power supply, (d) proper voltage and current availability at the field device at the farthest end, (e) device configuration checking and their signal analysis, (f) terminators at either end of the segment, and (g) cable insulation and capacitance.

23.13.1   FOUNDATION FIELDBUS DEVICE COMMISSIONING

A Foundation Fieldbus device is identified by (a) device ID, (b) device name (TAG), and (c) device address. The device ID is a 32-byte device identifier. It is a hardware address that is totally unique and unambiguously distinguishes one device from the other. The manufacturer sets this 32 character identifier. A device ID identifies the manufacturer, serial number of the device, model, etc. The device name (TAG) is set by the user. It identifies the function of the device and its physical location in the process. The device address is a unique one and is automatically set by the system. A device is associated to its particular configuration by correlating the device configuration tag with its device ID.

Common users interact with fieldbus devices based on tags although the node address is shown in the live list. A device is detected within seconds of its connection to the network. It is possible to connect/disconnect a device to the network without disturbing the same.

23.13.2   Profibus PA Fieldbus Device Commissioning

For PROFIBUS PA, the device address range is 0–126 and is unique in nature. Care must be taken to avoid address duplication, otherwise the system would lose control over the network. Thus, tag and address cross references are done to ensure avoidance of address duplication.

Address for each device must be set correctly for proper device identification. The address may be set locally for each device or remotely by the device configuration tool. For local setting, internal hardware DIP switches are used or externally by means of a local digital display. A device which is not on the operational network can be set remotely.

23.14   HOST COMMISSIONING

An interface or a linking device connects the host to field-level networks. The linking device or the interface may have multiple ports, allowing several networks to be connected to such an interface. Device support files from different manufacturers have to be loaded in the host so that it can interact with the devices placed in the field. The configuration tool has a dedicated folder for support files with subfolders for each type of device and manufacturers. The support files can be downloaded from the manufacturer's website so that configuration work can be completed even before the arrival of field devices at the site. There are several standard support files that are required to be installed for proper system operation.

23.15   ADDRESSING VIA ETHERNET

Data transfer between two nodes in a network takes place via transmission control protocol (TCP) and user datagram protocol (UDP), while Internet protocol (IP) is used for networking purpose. Ethernet is used for wiring LANs. Wiring and hardware access in FOUNDATION Fieldbus HSE, PROFInet, and Modbus/TCP are based on Ethernet IEEE 802.3/ ISO 8022.

On a network, Ethernet devices with different protocols can coexist without conflict. This is because they are built on the same platform, but in such cases they cannot talk to each other due to their differing user layers.

Addressing in Ethernet is done via Ethernet MAC address. This is a unique address set in hardware by the manufacturer which ensures that no two identical MAC addresses exist and eliminates any chance of conflict. This unique MAC address has a centrally administered code along with a unique identifier. On the other hand, the IP network address can be set either by the user or else assigned automatically.

23.16 ETHERNET

A LAN is a conglomeration of several computers connected to a network in a localized geographical area. LANs can work in isolation catering to the needs of an organization. Today's LANs are mostly linked to a wide area network (WAN) or the Internet.

Ethernet, token ring, token bus, FDDI, ATM LAN technologies are used for LAN. Out of these, Ethernet technology has become the most dominant.

23.16.1 IEEE ETHERNET STANDARDS

IEEE 802.3 committee has developed standards for the different Ethernet versions available.

Ethernet has travelled through several generations with the main concept remaining unchanged but has adapted to new technologies and the changing market needs.

The IEEE standard (called Project 802) for LAN is shown in Figure 23.20, along with the traditional OSI model. In this standard, the data link layer is divided into two sub layers—logical link control (LLC) and the MAC, with the former residing over the latter. LLC provides flow and error control for the upper layer protocols. It also supports a part of the framing actions. Framing is jointly supported by both LLC and MAC. It may however be mentioned that most upper layer protocols, including IP, do not use LLC in their OSI formatting.

23.16.2 TOPOLOGIES

Ethernet mostly uses star topology in which each node is connected to its own port on the central hub. Each segment is totally separate and is connected to the hub like the spokes in a cycle. Even if a segment fails, normal operation is not hampered and segments can be connected or disconnected under power.

Upper layers	Upper layers			
Data link layer	LLC			
	Ethernet MAC	Token ring MAC	Token bus MAC	- - - - - -
Physical layer	Ethernet physical layers	Token ring physical layer	Token bus physical layer	- - - - - -
OSI or Internet model	IEEE standard			

FIGURE 23.20 The IEEE standard for LAN and its comparison with OSI. (Courtesy: B. A. Forouzan. Data Communications and Networking, 4th Edition, Special Indian Edition, TMH Companies Inc., New Delhi, India, p. 396, 2006.)

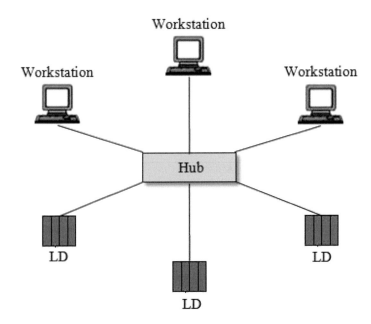

FIGURE 23.21   The star topology. (Courtesy: J. Berge. Fieldbuses for Process Control: Engineering, Operation and Maintenance, ISA, Research Triangle Park, NC, p. 116, 2004.)

Such a star connection topology is shown in Figure 23.21. There is a limit to the segment length due to signal attenuation along the wires.

When several star connected hubs are connected by means of a hub, a tree formation results, shown in Figure 23.22. In tree, branch, or star topology, disconnection of a device does not impact other devices while it is so in case of daisy chain topology.

23.17   IP BASICS

In fieldbus-based process control systems, linking devices, servers, workstations, etc. belong to the host level, while FOUNDATION Fieldbus HSE and PROFInet belong to IP. The IP addresses in FF HSE and PROFInet devices are usually configured by network administrator.

The total IP network address consists of two parts: the first part represents which network node is on and the second part represents network node address. Each of the host networks has an IP address. In IP version 4, the address is 4 bytes, each separated by a dot. These numbers are assigned by IANA (Internet Assigned Number Authority). It has set aside a three blocks of IP addresses for private use, exactly alike the ISM band. The process control system can be thought of as a private network which does not require it to be permanently connected to the Internet. In a way it is a

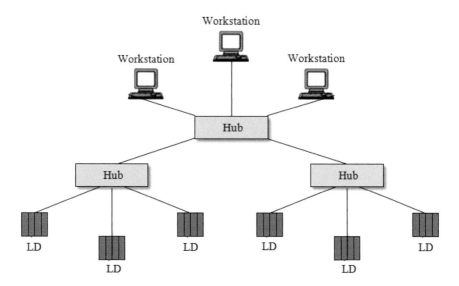

FIGURE 23.22 A tree topology resulting from several star connections. (Courtesy: J. Berge. Fieldbuses for Process Control: Engineering, Operation and Maintenance, ISA, Research Triangle Park, NC, p. 116, 2004.)

private network and uses the three free IP network addresses for the host-level networks.

Routers, which have already been discussed in Section 23.3.2.6, forward packets of information based on the IP network addresses. Locally, at the host level, the switches and bridges take care of Ethernet MAC address. Because both FF HSE and PROFInet use the Internet, hence they both use IP routers for transporting information across the Internet, i.e., between networks.

23.18   IP COMMISSIONING

IP network addresses can be configured manually even before a device is connected to the network. It can also be assigned automatically by dynamic host configuration tool (DHCP) server.

A Ping command checks the proper operation of a device after it is connected to the network. If the Ping times out, several possibilities arise. These are: the device is not wired or operating properly, it is connected to the wrong network, or the network parameters are configured wrongly. The Ping command can be used to confirm whether an IP address is already in use or not.

It should be noted at this point that IP routers are used by FF HSE and PROFInet to reach out to the destination network node while the Ethernet MAC address is used by switches and bridges to identify the particular device on a local network.

23.18.1 SUBNET

Help of IP routers are taken to divide a large network into smaller ones to counter the effect of huge broadcast network traffic. Likewise, subnet (sub network) is used to divide the local network into smaller segments having fewer devices in each. In the subnet mask, a portion is dedicated for identifying the IP part of the network while the rest portion is meant for device address identification. A device is always identified with a particular subnet mask.

23.19 MANUAL IP CONFIGURATION

Configuration involves setting the subnet mask and also the router IP address in the devices and workstations. This IP address is called the Default Gateway. Manual configuration involves simply keying in the address information. The addresses of either the devices or that of the network nodes must be unique. Address duplication must be avoided. Inadvertent address duplication would result in failed communication and disrupt supervision and control. To avoid the above, it is suggested to have address and its corresponding tag cross reference.

A manual addressing mode is static in nature. A device, that does not have a local interface, must be connected to a computer on a separate network that is running a special configuration tool.

23.20 AUTOMATIC IP CONFIGURATION

When configuring a large network, it is advisable to configure the system automatically because it is the best option. A DHCP server helps in automatic address assignment by simply connecting the devices to the network. The DHCP server assigns unique IP addresses to the nodes in the network. This addressing is dynamic because they may change with time.

24 Wireless Sensor Networks

24.1 INTRODUCTION

Wireless communication in the areas of process and manufacturing industries has already emerged as a strong contender compared to its wired counterpart. Advancements in wireless networking technology, particularly in the short haul communication, offers a tremendous opportunity for networking field devices with the host stations. The process measurement, control, and communication field is steadfastly moving toward the wireless strategy as the initial shortcomings are being removed and as the system is becoming more and more rugged and dependable. For any process control system to be networked wirelessly, factors which must be taken into consideration are: reliability, scalability, and real-time mixed data transmission, availability, security and coexistence with other networks. Some other issues are: sensor type, its location, power consumption, and local data processing ability and time stamping.

Data transmission by wireless means is a preferred mode because laying of cables poses a serious challenge. Each sensor in the field behaves as a node. The networked sensors sense field (process) data, compute, and process the same locally before being sent to the sink or gateway by multi-hop communication by following established protocols. The sensor nodes behave as transreceivers—i.e., they both act as data originators and data routers. Local data processing leads to reduced load on the transmitting medium.

A typical sensor network arrangement, shown in Figure 24.1, consists of sensor nodes, sink or gateway, Internet/satellite, and task manager before reaching the end user.

Finally, a wireless network may be personal area network (PAN), local area network (LAN), metropolitan area network (MAN), wide area network (WAN), or global area network (GAN) type.

24.2 WIRED VERSUS WIRELESS: A COMPARISON

A wired network may face difficulty on many counts like: trouble shooting, particularly loose connections, higher inventory, more manpower, susceptible to disconnections, cabling difficulty during plant expansions,

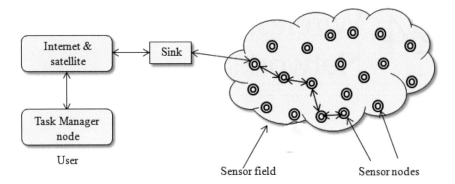

FIGURE 24.1   A typical wireless sensor network.

susceptibility to lightning surges, difficulty in maintaining a cabling network, and a hostile and geographically huge network.

Wireless networks can easily be expanded during plant expansions, reduces blind spots leading to higher yields, better plant surveillance. Some issues remain, though, for wireless networks like reliability, security, interference, cross-talk from neighboring channels, jamming the transmission by using the same frequency in the vicinity for malicious purposes as also interception and eaves dropping.

24.3   ISM BAND

Wireless sensor networks (WSNs) use the license free ISM (Industrial, Scientific, and Medical) band for error free communication. ISM is an unlicensed band having three ranges: 902–928 MHz, 2.4–2.835 GHz, and 5.725–5.85 GHz, respectively. Initially, medical diathermy, radio frequency (RF) heating, microwave ovens used the ISM band.

When two or more wireless devices operate in the vicinity of each other in the same band, electromagnetic radiation from one may affect the operation of the other. Design of each network should be such that the above may not occur. Now-a-days widespread use of wireless networks are increasingly being used in areas of: wireless computer networks using LANs, bluetooth and near field devices, cordless phones, etc.

Figure 24.2 shows the license free electromagnetic spectrum. A band starting at 863 MHz is used in Europe. The different frequency bands have different output powers, bandwidths, and duty cycles.

24.4   WIRELESS STANDARDS

For different needs and purposes, different standard wireless protocols are used to send data/messages. Some of the standards are: IEEE 802.11, IEEE

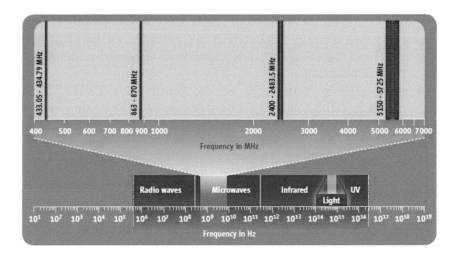

FIGURE 24.2 License free frequency bands.

802.15, IEEE 802.16, etc. These standards are again having several subdivisions within them. Wireless protocol to be used for a particular application depends on the specific needs of the same.

While Wi-Fi is a LAN technology based on IEEE 802.11 specifications, sometimes also called wireless Ethernet and introduced in 1977, WiMAX is based on IEEE 802.16. Wi-Fi uses carrier sense multiple access with collision avoidance (CSMA-CD) as the medium access control (MAC) protocol and a half duplex shared media configuration. WiMAX is better in sensor networking applications because of its better coverage and cell capacity. WiMAX uses full duplex transmission mode for data transmission and a request/grant mechanism that assumes separate channels for inbound and outbound transmissions.

Bluetooth, sometimes called small wireless LANs, is a low power short-range open wireless communication protocol. It uses IEEE 802.15.1 and a channel hopping scheme ensuring low latency and high throughput. The hopping frequency of transreceivers is 1600 hops per second and thus reduces chances of interference and fading to a minimum. Bluetooth uses Time Division Duplex Time Division Multiple Access (TDD-TDMA) scheme for transmission of data or message. Separate hops are used for to and fro communication. It does not guarantee end-to-end communication delay. It uses a half-duplex method for transmission.

Like Bluetooth, ZigBee uses PAN technology. Power associated with ZigBee is ultra-low and is used in monitoring and control. It is based on IEEE 802.15.4 physical and MAC layers. It is a low-power, low-cost, low data rate communication protocol. ZigBee, like Bluetooth, does not guarantee end-to-end communication delay. It uses direct sequence spread spectrum (DSSS) for transmission of information. It does not provide frequency and path diversity. ZigBee does not guarantee reliability against interference and obstacles. Thus, it is seldom used in industries, although it is a secure communication.

Wireless highway addressable remote transmission (WHART) is based on IEEE 802.15.4 and is backward compatible with existing HART devices. It uses 2.4 GHz ISM unlicensed frequency band for communication and is the first open standard protocol for WSNs in the field of automation and control. It uses TDMA technology and has 250 kbits per second data rate. It supports channel hopping with a 5 MHz separation between any two adjacent channels. Security in WHART is provided by MAC layer and network layer through a 128 AES algorithm. Frequency diversity, path diversity, and proper message delivery methods ensure high reliability in networks that employ WHART.

ISA 100.11a, like WHART, is based on IEEE 802.15.4 and uses the unlicensed 2.4 GHz ISM band. It is not backward compatible. Some of the features associated with this protocol are: low cost, less complexity, low power consumption, robustness to RF interference, interoperability, scalability. It operates in both star and mesh topology. Message protection is ensured by using a 128 AES algorithm that includes asymmetric cryptography. TDMA mode is used in ISA 100.11a ensuring sleeping routers and thus power consumption is very low.

24.5   STRUCTURE OF A NODE

A sensor node consists of several modules like sensing unit, processing unit, power unit, and a communication unit. It is shown in Figure 24.3.

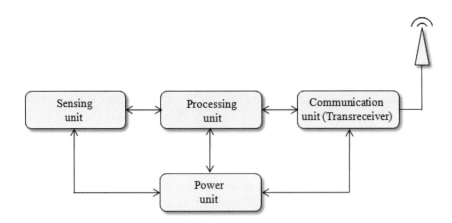

FIGURE 24.3   Basic building blocks of a sensor node.

A sensing unit may consist of a single sensor or multiple ones. Physical parameters like temperature, pressure, etc. are normally sensed by the sensor unit. It is then converted into digital signal by an ADC and fed to the processing unit.

The processing unit accepts data from the output of ADC and interfaces it to the physical radio layer, manages radio network protocol. By software control, this unit also manages to minimize power consumptions in radio subsystems, sensors, signal conditioning, and communications. It is preprogrammed to perform tasks assigned to it. Performance of the processing unit is evaluated on the basis of data rate, processing speed, memory, and peripherals.

The communication unit consists of a common transceiver which connects the nodes to the network. It is mainly used to transmit and receive data/information among the various nodes and base station and vice versa. Mainly, there are four states in a communication unit: transmit, receive, idle, and sleep.

A wireless sensor network is an infrastructure less power constrained network and supplies power to the node. Power is drawn from rechargeable batteries, or solar power which is a better option for inhospitable terrains.

A sensor node performs data acquisition while processing unit performs analysis, aggregation, compression, and fusion tasks. Another block within the node performs management, coordination, and configuration of data.

24.6   A NETWORK OF SENSORS

A sensor network, arranged as shown in Figure 24.4, contains remote sensor, clustering node (or intermediate processing node), and a final processing node. The sensors are connected in single hop or multi-hop manner. The whole combination is termed as a sensor field.

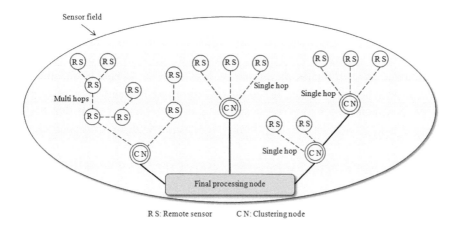

FIGURE 24.4   A typical sensor network arrangement.

It operates reliably if both the location of sensors as also their time stampings are done properly. The nodes communicate with the sink. The nodes must have location and positioning knowledge with the help of global positioning system (GPS). The nodes collect data, analyze, and compress the same and forward (route) it to the sink.

Data is collected at the base station (final processing node) and hence nodes closer to the base station will have more data burden. A judicious routing mechanism will alleviate this problem substantially.

24.7 FEATURES OF A WIRELESS SENSOR NETWORK

A wireless sensor network is scalable and uses unlicensed ISM band for data transmission purpose. It is an ad hoc network that organizes into an infrastructure less network. The topology of such a network vary frequently and communication is many to one (data is collected at a base station) rather than peer-to-peer type. Some nodes are densely populated while some others are otherwise. The nodes are power constrained and self configuring in nature. Nodes in a WSN are power constrained and robust to topology changes like addition/withdrawal of a node. Data from several groups of nodes are clustered and hence fewer transmissions are needed.

24.8 CHALLENGES IN A WIRELESS SENSOR NETWORK

Since a wireless sensor network is very much power constrained, hence communication protocols for such a network must have very high power efficiency. Thus, such protocols mainly focus on power consumption and efficiency, its traditional counterpart mainly deals with throughput and delay.

Node deployment in WSNs is another key area on which protocol development largely depends. Random deployment of sensor nodes needs the development of protocols which must be self-organizing. Such nodes should collaborate with other nodes and can adapt to failures without any human intervention. In a majority of cases, nodes deployed in WSNs are left unattended and as such repair, maintenance, and adaption to new environment should be inherent in them. It should be seen that such self-managed features of sensor nodes must be designed and implemented such that they operate over a considerable length of time with limited power resources.

24.9 CONNECTIVITY CONSTRAINTS INTO INTERNET

Energy and memory efficiency are two performance metrics which are not of much interest in case of normal protocols, while it is just the reverse for wireless sensor networks. Hence for seamless connectivity to Internet, a wireless sensor network will require significant modifications in design.

As wireless technology is still evolving and increasingly being applied across various sectors, different network architectures for catering to different fields are used such as cognitive radio networks, mesh networks, wireless sensor networks, etc. In order to integrate these various wireless networks with Internet requires interoperability. It is vital to develop location and spectrum aware cross layer communication protocols as well as heterogeneous network management tools for seamless connectivity of these protocols with the Internet. The 6LoWPAN standard was developed to integrate IPv6 standard with low power sensor nodes deployed in WSNs. To integrate an IPv6 based device and a sensor mote, the packet header of IPv6 is compressed to the extent that it fits into properly to the sensor mote. Since both WSNs and wireless LAN (WLAN), operate in the same spectrum range, their coexistence at the MAC layer poses a major challenge.

Other major issues which need to be addressed are: (a) end-to-end routing (or connectivity) between a sensor node and Internet, (b) existing transport layer solutions for WSNs are incompatible with transmission control protocol (TCP) and user datagram protocol (UDP) which are extensively used in Internet service. Again, transport protocols for WSNs should be so designed that a seamless reliable transport of sensor data and other event features are guaranteed throughout the sensor network.

24.10 TOPOLOGY

The manner in which sensor nodes in a network are interconnected is called topology. The different topologies employed in wireless network are: star (single point-to-multipoint), mesh, and star-mesh (hybrid).

A star network is shown in Figure 24.5. The base station can receive/send data from/to a remote node. A remote node cannot send data to another

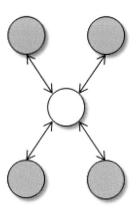

FIGURE 24.5 A star network.

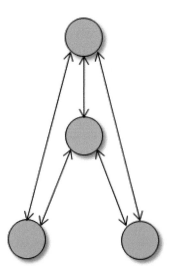

FIGURE 24.6 A mesh network.

remote node, i.e., peer-to-peer communication is not possible. Advantages of a star network include its simplicity, fastness, and a remote node's power consumption is kept to a minimum. Latency between the base station and a remote node remains low always. But the base station must be within the radio transmission range of all the nodes connected to it. This topology is also weak to certain extent because if the base station becomes out of order, this part of the network goes offline.

A mesh network, shown in Figure 24.6, is used when the sensors are placed over a large geographic area having high redundancy. Since WSN is a self-configuring network, it automatically determines the best path if a sensor en route suddenly fails. This topology has the advantage of both redundancy and scalability.

Mesh networks are used when a node wants to send data to a node which is outside its radio transmission range. In that case, the node transmits its data to a neighboring node which is within its radio transmission range. From the second node, it is passed on to a third node. The process is repeated until the message reaches its final destination. This is called multi-hop communication.

Because of multi-hop type of communication in a mesh network, the nodes which lie in between and are required to pass and forward the information, will consume more battery power thereby limiting their lives. Another associated disadvantage of multi-hop communication is the time of delivery of a message—it becomes more as number of hops become more for a message to reach its destination.

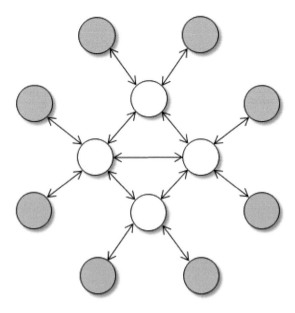

FIGURE 24.7   A star-mesh or hybrid network.

Multi-hop communication undertaken in mesh topology results in less distance covered in a single hop. Since signal path loss is inverse exponent of path distance, hence quality of signal received at the sink is ensured.

A star-mesh or hybrid topology is shown in Figure 24.7. It derives the advantages of both the star and the mesh networks, i.e., reduced power consumption of a star network and self repairing capability of mesh network.

24.11   COEXISTENCE ISSUES

When several wireless channels are simultaneously sending data in the vicinity of each other, RF interference becomes an issue which may affect correct reception of data at the receiver. Coexistence is defined as "the ability of one system to perform a task in a given environment where other systems have an ability to perform their tasks and may or may not be using the same set of rules" (Industrial Automation Technologies, Chanchal Dey and Sunit Kumar Sen, CRC Press, 2020). As an example, IEEE 802.15.4 and IEEE 802.11b/g both use the 2.4 GHz unlicensed frequency band for data transmission. If these two protocols are employed for data transmission in the vicinity of each other, RF interference may affect both channels. Care must be taken to reduce this interference to improve the quality of reception. The problem is aggravated when two messages using these protocols and having sufficient energy collide or overlap with each other in time or frequency domain.

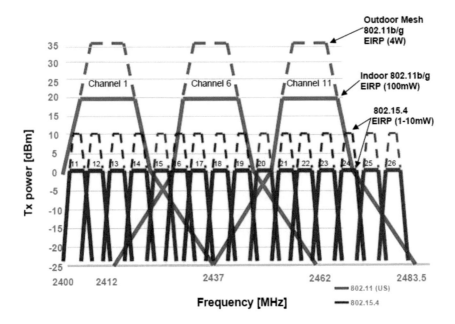

FIGURE 24.8 Response of 802.15.4 and 802.11 b/g in 2.4 GHz ISM band.

Several techniques are adopted to address the above issue which are: time diversity, frequency diversity, power diversity, coding diversity, space diversity, blacklisting, and channel assessment.

Figure 24.8 shows the response of 802.15.4 and 802.11 b/g in 2.4 GHz ISM band. As can be seen from the figure, the former has 16 channels while the latter has 3. Channel numbers 15, 20, 25, and 26 (for North America) or 15, 16, 21, and 22 (for Europe) of 802.15.4 will be less influenced from the side slopes of 802.11 b/g as is evident from the figure. WHART utilizes a pseudo-random channel hopping sequence to reduce this interference by using these non-overlapping channels.

WHART uses TDMA technology to send one message per frequency channel at any given instant to avoid collision. WHART networks can be configured to avoid certain channels which are also very much used by other networks. This way collision and interference can be avoided. WHART networks listen to a channel before initiating message/data transmission. If the channel is busy, the current transmission is set aside and a future slot is allocated for its retransmission.

In coding diversity, DSSS technique is used to spread a message over the entire bandwidth of the selected channel. This reduces interference.

Space diversity is achieved by employing mesh networks. In this, the originating message may be delivered to the gateway via different paths—depending on traffic load and the availability of intermediate nodes.

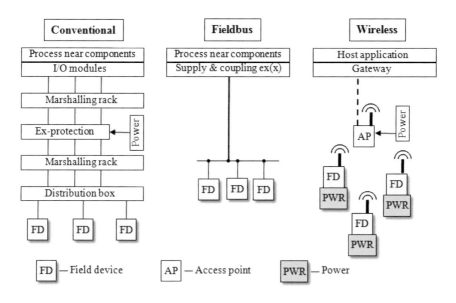

FIGURE 24.9 Conventional, fieldbus, and wireless network architectures: A comparison. (Courtesy: W. Ikram and N. F. Thornhill, Wireless Communication in Process Automation: A Survey of Opportunities, Requirements, Concerns and Challenges, Control 2010, Coventry, UK 2010,)

24.12   CONVENTIONAL, FIELDBUS AND WIRELESS NETWORK ARCHITECTURE: A COMPARISON

Architectures of a conventional, fieldbus system and a wireless sensor network is shown in Figure 24.9.

The major advantage of a wireless network is the absence of wired connections prevalent in other systems. It leads to less maintenance, lesser cost, less manpower deployment, etc. But a wireless network should be protected against eavesdropping, RF interference from neighboring channels, and that it must stay connected irrespective of static or mobile sensors.

24.13   PROTOCOLS IN WIRELESS SENSOR NETWORKS

While communication in a wired network takes place over a guided medium, the same takes place via electromagnetic signal transmission through the air for wireless networks. For the latter case, the sensor nodes need to self-organize to have multi-hop communication facility. For higher efficiency, the nodes must share all the resources in a very efficient manner. This can be achieved by an efficient MAC that determines the overall performance of the network.

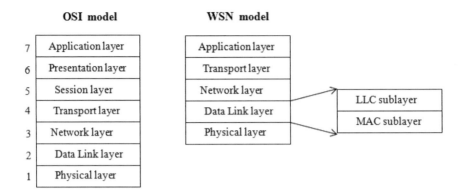

FIGURE 24.10 Protocol stack of a wireless sensor network.

The protocol stack used by the sensor nodes and the sink is shown in Figure 24.10. Since many nodes are used in the network, it is to be ensured that a single node gets access to the transmission medium at any point of time.

The data link layer (DLL) is subdivided into two sublayers: the upper logical link control (LLC) sublayer and the lower MAC sublayer. LLC is required to accommodate the logic required for shared access of the medium and to support different MAC versions. The different versions are required depending on the topology of the sensor nodes, the quality of service (QoS) required and the characteristics of the communication channel.

Functions of the MAC sublayer include: (a) regulating access to the shared medium and (b) a header field, comprising the address information, is appended at the front of a data stream and a trailer field at the tail of the same for error detection purposes. (c) At the receiver, address and error control fields are removed and decision taken regarding correctness of both address and data received.

To decide upon which node can have access to the communication channel, the nodes must exchange some information among themselves. Thus, complexity and overhead of the MAC protocol increases.

Performance metrics of the protocol depends on a host of factors like: throughput, delay, robustness, stability, scalability, fairness, and energy efficiency. A higher throughput makes the system fast resulting in less delay for a message to reach the receiver. The network is said to be robust if it is immune to errors and misinformation. Stability refers to the ability to handle fluctuations in traffic load over a period of time in a reliable manner. A system is said to be scalable if its performance remains invariant with the size of the network. Fairness is related to channel capacity being equitably shared among the competing nodes, without degrading throughput. Energy efficiency is one of the most important issues in a wireless network and is particularly so when the network is geographically widely dispersed. Some factors that lead to low energy efficiency are: collision, control packet overhead, idle listening, overhearing, and switching between different modes.

Performance of the network depends on the choice of the MAC protocol. Several such protocols exist which are: fixed assignment protocol, demand assignment protocol, random assignment protocol, flooding, and gossiping. They have been developed to strike a balance between achieving judicious resource allocation and overhead needed to implement the resources.

24.14 SECURITY

Resource-constrained wireless sensor nodes are easy targets from external malicious attacks like phishing, etc. An adversary can listen to the traffic flow in the network, manipulate data, or even impersonate as one of the operating nodes.

Normally, such networks have huge number of nodes and providing security to each and every node is practically challenging. Addition of nodes due to expansion or removal of a node which suddenly becomes inoperative poses a challenge to software design engineers.

Cryptography ensures a secure data path between a transmitter and a receiver. In this, an ordinary or plain text is converted into cipher text, called encryption, at the transmitter. The reverse process, called decryption is undertaken at the receiver. Data encryption requires an algorithm and a key. While the algorithm may not be secret, total secrecy is maintained about the key used for encryption. The number of bits comprising a key is huge and it is just impossible to get the composition of the key. Two types of cryptography are employed: symmetric and asymmetric. In a symmetric cryptosystem, the same key is used for encryption and decryption, while separate keys are used in asymmetric cryptosystem. Cryptography ensures data integrity, confidentiality, device and message authentication. The security strength of any cryptosystem depends on the key used and not on the algorithm.

Security services also include validation, access control, scalability, and data freshness. Validation implies correctness of authorization to use resources, while access control refers to restricting access to resources. Scalability refers to addition of nodes and faithful operation of the system without compromising security. Data freshness implies that message remains fresh, i.e., they are not reused and also they are in order.

An efficient key management scheme (KMS) is a must so that data security or phishing attacks can be dealt with strongly. When a node is added to or deleted from the system, the KMS plays a vital role for a secure and reliable network environment.

Some of the security goals in a sensor network are: confidentiality, integrity, authentication, and availability. Confidentiality implies the ability of the system to guard the message from eavesdroppers as the message successively passes from one node to the next one till it reaches the receiver. Integrity refers to the message not being tampered/changed/altered while it moves on in the network. Authentication refers to the ability to confirm about the origin of a message. Availability implies ensuring that the network is available for a message to move on in the network.

Freshness of data becomes a serious security issue particularly when WSN nodes use shared keys for message communication. In such cases an adversary can launch a replay attack using the old key as the new key is being refreshed and propagated to all the nodes connected in the network. Data freshness can be ensured by adding nonce or time stamp to each data packet.

25 WirelessHART (WHART)

25.1 INTRODUCTION

WirelessHART (or WHART in short), like ISA100.11a, is fast becoming a wireless emerging standard in industrial wireless communication arena which promises to revolutionize the process control and automation industry. Existing wireless technologies like Bluetooth, ZigBee, or Wi-Fi cannot be used in industrial plants because of their inherent drawbacks. Bluetooth is primarily targeted at personal area networks (PAN) whose range is limited to within 10 m. Furthermore Bluetooth supports only star type network topology with a master having a maximum of 7 slaves largely limiting its application in process industries. Although ZigBee supports direct sequence spread spectrum (DSSS), its performance is severely degraded and affected in presence of noise normally encountered in industries. Wi-Fi is not at all a good choice for industrial communication because of its inability to support channel hopping. Power consumption is another deterring factor that goes against its possible use in industries.

WHART is the first open wireless communication standard specifically designed for process industries. It was officially released in September 2007 as part of the HART specification and was a part of IEC 61158. WHART specification was approved by IEC as an international wireless standard IEC 62591 Ed. 1.0 for wireless communication and process automation in March, 2010.

WHART was developed keeping in mind industrial applications like device status and diagnostics monitoring, calibration, critical data monitoring, troubleshooting, commissioning, and supervisory process control. WHART provides robust and reliable communication using DSSS channel hopping scheme based on IEEE 802.15.4, mesh (redundant data paths), and retry mechanisms.

Wireless mesh networks (WMNs) are gaining more and more market acceptance because of advancements in wireless technology with multi-input multi-output systems and smart antennas. Some of the key features of WMNs are: self-configured and self-organized, a mix of wired and wireless devices, every network device acts as a router so that the nodes can communicate with each other through multi-hopping, uses gateway to communicate between devices, etc.

25.2   KEY FEATURES

WirelessHART is a highly reliable, self-healing, self-organizing, time synchronized redundant path wireless mesh network communication protocol designed to meet process industry requirements. It is backward compatible with existing HART devices, i.e., it supports HART command structure and device description language (DLL). WirelessHART operates in the 2.4 GHz license free ISM radio band utilizing IEEE 802.15.4 compatible DSSS radios with channel hopping facility on a packet-to-packet basis.

It uses time division multiple access (TDMA) technology to arbitrate and coordinate communications between various devices connected to the network. It supports multi-messaging modes that include one way publishing of process, auto-segmented block transfers of large data sets and has facilities like spontaneous notification of exceptions and also ad-hoc request/response.

It provides highly secure centrally managed communications using advanced encryption standard (AES)-128 block ciphers with individual join and session keys and data link level network key. The security services are provided by media access control (MAC) layer and network layer.

WirelessHART is a hybrid network consisting of wireless and wired devices. It has channel *blacklisting* facility which disallows use of determined channels. Use of redundant paths and redundant communications leads to increased reliability.

By using TDMA for medium access reduces probability of collisions. Also power consumption is drastically reduced by ensuring radio transmitters remain awake in the allotted slots only by prescheduling the network manager.

WirelessHART employs 15 channels in the ISM 2.4 GHz spectrum. Channel selection is based on absolute slot number (ASN). The slot numbers are unique and monotonically increase. Slots and superframes are assigned by the network manager. Slots can be keep-alive, join request-response, ad-hoc request-response or special purpose types. Pseudo random channel usage by employing channel hopping technique ensures that interference on one or several channels does not prevent reliable communication. It frequently changes channel for transfer of packets for reliable data delivery. Data traffic can be periodic or sporadic. The field devices of WHART must have mandatory routing capability while the adapter connects to wired HART network. The access points provide access to the wireless network.

There is a single network manager which is responsible for configuring the network as also scheduling and routing of data packets. The security manager is responsible for management and distribution of keys.

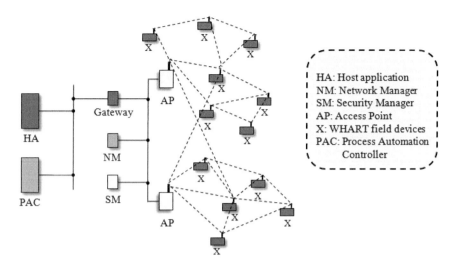

FIGURE 25.1 A WHART network architecture.

25.3   WHART NETWORK ARCHITECTURE

The architecture of a WHART network is shown in Figure 25.1. It consists of WHART field devices, HART-enabled field devices, handheld field devices, access points, gateway, adapters, network manager, security manager, etc.

Communication between the wireless field devices and the wired host system is enabled by gateway. Point-to-point and mesh topology are used to connect WHART field devices. The network manager provides new schedules and routing information as new devices continue to join the network. For any wireless device, there should be at least two connected neighbors that can route the traffic using graph routing. Use of mesh topology along with frequency hopping and TDMA for communication makes the system robust and reliable in process control applications.

25.4   PROTOCOL STACK

The protocol stack of WHART is shown in Figure 25.2, along with open systems interconnection (OSI) protocol. WHART protocol stack consists of five layers: physical layer, data link layer, network layer, transport layer, and application layer. In addition, a central network manager manages the routing and arbitration of communication schedules.

25.4.1   PHYSICAL LAYER

Physical layer is based on IEEE 802.15.4-2006 with a data rate of 250 kbps (62.5 kbaud) having an operating frequency of 2400–2483.5 MHz.

Application layer	Command oriented, predefined data types and applications
Presentation layer	
Session layer	
Transport layer	Auto segmented transfer of large data sets, reliable stream transport and negotiated segment sizes
Network layer	Power optimized, redundant path, self healing and mesh network
Data link layer	Secure, reliable time synchronized TDMA/CSMA , frequency agile ARQ
Physical layer	2.4 GHz, wireless, 802.15.4 based radios, 10 dBm T_x power

FIGURE 25.2 WHART protocol stack.

Its channels are numbered from 11 to 25, with a 5 MHz gap between two adjacent channels. Modulation used is O-QPSK with DSSS. The nominal transmit power is 10 dBm which is adjustable in discrete steps. The physical layer protocol data units (PDU) is IEEE compliant with a maximum payload of 127 bytes. Channel hopping takes place on a packet by packet basis. It has clear channel assessment (CCA) facility by which it first checks whether a channel is free before sending a data packet.

The physical layer PDU (PPDU) is shown in Figure 25.3. It consists of a preamble of 4 bytes, a delimiter of 1 byte, a PPDU of length 1byte and a variable length payload.

25.4.2 DATA LINK LAYER

DLL establishes mechanism for reliable error-free and secure communication between devices to occur in distinct time slots. To ensure error-free communication, WHART DLL introduces channel hopping with channel blacklisting facility using TDMA technology. TDMA uses precise time slots between which secure and deterministic communication takes place. The data link layer PDU (DLPDU) is ciphered by using a 128-bit AES

Preamble	Delimiter	Length	PPDU payload

FIGURE 25.3 The physical layer PDU.

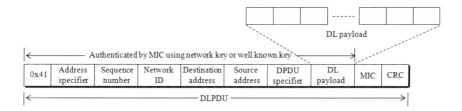

FIGURE 25.4 The data link layer PDU.

cipher algorithm for achieving secure communication at this layer. The data link layer PDU is shown in Figure 25.4.

The DLPDU consists of several fields which are self explanatory. The DLPDU packet type defines the contents of the DLPDU payload.

A series of 100 time slots per second form a superframe and WHART consists of multiple superframes. Figure 25.5 shows such superframes with time slots. One superframe is always enabled at a time. To ensure contention free access to the medium, a time slot is allotted to two devices at the same time—one as the source and other as the destination. The length of a superframe is fixed when it is active but can vary or be modified when inactive. Thus, different superframes may have different lengths.

A management superframe has 6400 slots. The slots in such a superframe are used for join request/response; ad-hoc request/response; keep alive request or special purpose slots used for block transfers or handling requests from handhelds.

In a time slot, a DLPDU is transmitted from the source, followed by an ACK DLPDU from the destination. If the destination receives the DLPDU from the source successfully, only then the ACK DLPDU will be transmitted by the destination. Again, if the source does not successfully receive and validate the ACK DLPDU from the destination, transmission would be regarded as invalid and the DLPDU will be retransmitted by the source in the next available time slot. If a particular link fails to deliver data to a destination, repetitively, the link is discarded and data would be sent to

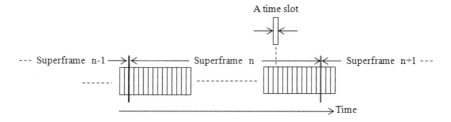

FIGURE 25.5 WHART slot time and superframes.

the destination via some alternate route based on the routing table of the source device.

When a source device is transmitting a DLPDU, the destination (RX) is in the listening mode, while when the destination sends back the acknowledgement ACK DLPDU, the source (TX) would be in the listening mode. The acknowledgement packet includes relevant timing information to continuously synchronize TDMA operation across the whole WHART network.

DLL addressing is based on IEEE 802.15.4 physical layer with a unique 64-bit IEEE extended address (IEEE EUI 64), a short 2 byte unique address within the network and a 2 byte network ID.

Time synchronization is maintained by the gateway which acts as the ultimate time source. The gateway synchronizes with the access points. Time synchronization is based on a tree-based clock adjustment procedure.

Data transfer on a data link layer can be periodic or occasional in nature. Again data transfer may require one slot for the first transmission, one slot for a possible retry on a separate channel or one slot for a second retry on another channel yet again.

25.4.3 Network Layer

Network layer performs several functions like: packet routing, encapsulating the transport layer message exchanged across the network, ensuring secure end-to-end communication, block data transfer, end-to-end security, acknowledging the broadcasts, etc.

The network manager is responsible for configuring the routing tables of every device in the network, scheduling and managing communications between different WHART devices. Every device in the network should be able to forward the incoming packets on behalf of other devices to support mesh technology. Three routing protocols defined in WHART: graph routing, source routing, and proxy routing.

- **Graph routing:** It is a collection of paths that connect the network nodes. The paths in each graph is created by the network manager and downloaded by each network device. Each device maintains a graph table with a list of graph IDs. That is each network device is preconfigured with graph table (information). When a packet is to be sent, the source device writes a specific graph ID (this is determined by the destination address) in the network header. When the device receives a packet containing the graph ID, information stored in the graph table is utilized to select the next path (hop).
- **Source routing:** In this, the source device includes in its header the entire route in the packet. Thus as the packet progresses through

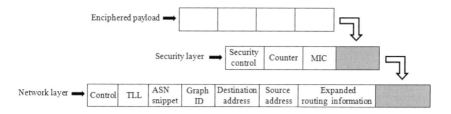

FIGURE 25.6   The WHART network layer PDU.

the nodes, each time the current node just forwards the packet to the next hop as per the information that is already embedded in the header of the packet. This way it goes on until the destination node is reached.

- **Proxy routing:** It is used when the device is yet to join the network.

The network layer PDU (NPDU) is shown in Figure 25.6. The frame begins with a control header that contains an addressing scheme employed and it also shows whether any special routes are used in the remainder of this byte. TTL stands for time-to-live and it is a counter which continues to decrement as the message passes through the hops one after the other. Thus, the counter value at any instant indicates how many more hops the message is to pass through before it reaches its destination. ASN snippet field provides performance information about the network. It specifies the time passed since a packet was formed. Graph ID routes the packet about the path the packet takes to reach the destination. The next two fields are destination and source addresses. Expanded routing information field indicates any additional routing information that may be needed to guide the packet to the destination.

The next field of the network layer is a security sublayer which has several fields and is responsible for data encryption and NPDU authentication. Security field indicates the type of security employed: join key, unicast session key, or broadcast session key. Data integrity is checked by message integrity code (MIC) field.

25.4.4 TRANSPORT LAYER

Figure 25.7 shows the transport layer PDU. It ensures end-to-end packet delivery for communications that require acknowledgement like request-response traffic. Data sets are automatically segmented at the source device and reassembled at the destination device.

Either the Join Key or any one of the session keys is used to encipher the TPDU. This layer ensures reliable communication. Devices which communicate with each other are provisioned with identical join keys. The

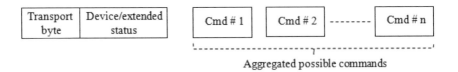

FIGURE 25.7   WHART transport layer PDU.

transport layer encapsulates the application layer data. It also acts as the convergence point between HART and WHART.

25.4.5   APPLICATION LAYER

WHART uses the standard HART application layer, which is command based and resides at the topmost layer. The commands are: universal commands, common practice commands, device family commands, and wireless commands. The universal commands are a set of commands which are supported by all WHART devices. The common practice commands are optional in nature and apply to a wide range of devices. The device-specific commands are manufacturer specific according to field device needs, the implementation of such commands are also optional in nature. The wireless commands support WHART products. All devices supporting WHART must implement these commands for proper network operation.

The application layer in WHART is responsible for parsing the message content, extracts the command number, executes the specific command and generating responses. The network manager utilizes the application layer commands to configure and manages the whole WHART network. Figure 25.8 shows the general format of the WHART application layer.

25.5   NETWORK COMPONENTS

Network components in a WHART network are network manager, security manager, virtual gateway, access points, host interface, adapters, routers, and field devices. Field devices are connected to the process or plant equipment. These are the sensors distributed throughout the plant which route and forward the data or information packets.

The concept of a distributed architecture with more than one access point is shown in Figure 25.9.

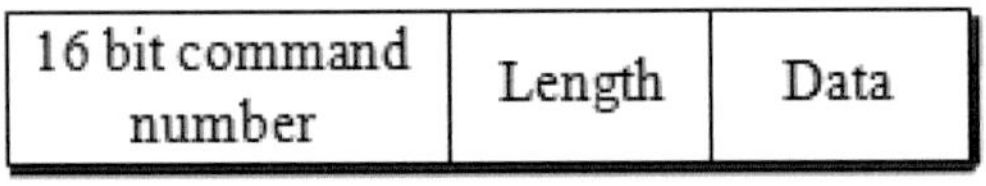

16 bit command number	Length	Data

FIGURE 25.8   The WHART application layer format.

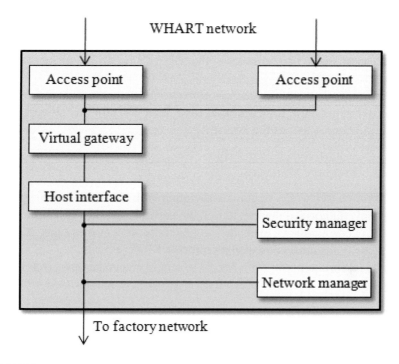

FIGURE 25.9   A distributed architecture of network components. (Courtesy: Wireless HART, Device Types-Gateways, HCF_LIT-119Rev. 1.0, 2010, Gerrit Lohmann-Pepperl+Fuchs, HART Communication Foundation, Austin, TX 78759, USA)

The access points shown in the figure receive and send the wireless messages. An access point receives data or messages from the virtual gateway and passes the same to the sensors via the host interface, taking the help of security and network manager. On the other hand, device data is put on the WHART network via the access point.

25.5.1   Network Manager

A network manager can be thought of as the receiver and distributor of HART commands. It is the control center of the whole network. The network manager is not a physical device but simply a software. It deals with forming, organizing, and monitoring the HART network. It configures the network, schedules required communication amongst devices, manages the routing in the network, and reports health of the network to the host application. In a network, there is only one active network manager at any point of time. The network manager receives the status of all the devices in the network. Based on this, it schedules communication between two devices. The network manager is responsible for managing dedicated and shared resources. When a device joins or leaves a network, it is the responsibility of the network manager to manage the same.

A measure of the performance of the WHART network manager can be gauged by

maximum number of devices connected to the network, time required to initialize the network, time required for a device to join the network, and the overall throughput of the system.

25.5.2 Security Manager

Security manager works in close association with the network manager. It allows only authorized devices to join the network and provides authorization and encryption keys to ensure proper encryption of messages. It manages security resources and also monitors network security.

Some important points with regard to security manager are given in the following:

- There is only one security manager for a WHART network.
- It can serve more than one network.
- The security manager and network manager work in a client-server fashion.
- Interface between the security manager and network manager is not defined in the WHART standard.

25.5.3 Gateway

A gateway enables communication and acts as a link between host applications and field devices in the network. It is functionally divided into a virtual gateway and one or more access points. There is one gateway per network. It is responsible for buffering, protocol conversion, and clock source. Conceptually, a gateway can be thought of as the wireless version of marshaling panels and junction boxes.

A gateway should have the following attributes:

- provides for network and security management functionalities.
- has multiple output protocols which would ensure proper integration to a range of different host applications like DCS, PLC, data historian, etc.
- should support multiple connections, i.e., it acts like a server. Thus, data can be sent to different end users.
- should be interoperable.
- should support secure transfer of all protocols over an Ethernet connection through robust encryption process.
- must have different security access for different users.

25.5.4 ADAPTER

An adapter allows existing HART field devices to be integrated into a WHART network. It provides a parallel communication path to the existing 4-20 mA current loop for the WHART network and provides wireless communication path to HART devices. An adapter should have a HART tag and must not affect normal 4-20 mA signal.

25.6 LATENCY AND JITTER

Time is required for measurement data to reach the controller input. This latency (time delay) and jitter (variation in latency) may affect proper control of the process.

WHART is a time synchronized control protocol with every field device. This synchronization is not available with most of the other protocols. This time synchronization is used to schedule measurements, almost eliminating latency and jitter.

Compared to some wired fieldbus systems, WHART is faster in nature. For example, in FOUNDATION Fieldbus, the transmission rate is 31.25 Kbits/s, while for WHART it is 250 Kbits/s. For the former, communication delay is 32 µs/bit and for the latter, it is only 4 µs/bit.

A time slot involves 10 ms for transmission and its acknowledgement to take place. A typical WHART message consisting of 128 bytes needs only 4 ms for the message to reach its destination. Thus, latency is absent in such a case.

Latency does not pose any problem as long as it is less than the process response time. For all practical purposes, this is so. Appropriate communication schedules, along with black listing of channels and channel hopping and proper networking with appropriately placed access points can reduce latency to such an extent that latency always remains much less than the response times of most of the physical processes encountered in practice.

25.7 COEXISTENCE TECHNIQUES

WHART operates in the license free ISM band which is also being used by other wireless systems like, Bluetooth, Wi-Fi, ZigBee, etc. Thus, any wireless industrial application has to confront signals coming from such systems. Efforts have been made so that WHART can coexist with these signals and can reliably communicate with other wireless field devices as also the wired host systems and applications. Several techniques have been adopted to achieve the above, which are discussed in the following subsections.

25.7.1 CHANNEL HOPPING

As already mentioned, WHART is a highly reliable, self-healing, self-organizing, time synchronized redundant path wireless mesh network. This self-organizing property of WHART ensures that the network can hop from one path to another in its passage to the receiver. The different paths are subject to varied amount of radio frequency (RF) interferences. Their effects are minimized by the self-organizing property of WHART networks.

WHART uses 15 channels having different frequencies in the ISM band. WHART uses a pseudo random channel hopping sequence to reduce chances of interference from other possible interfering systems such as Bluetooth, Wi-Fi, ZigBee, etc. This method ensures that any particular channel is not in use for any prolonged length of time, thus reducing any prolonged interference.

WHART utilizes one channel hopping scheme called, *slotted hopping*, in which pattern variation is done for different channels and shown in Figure 25.10. The channel hopping order is dependent on channel offset, ASN and also on the number of channels which are currently active.

25.7.2 DSSS

DSSS scheme provides an improved signal to noise level. It improves receiver sensitivity via digital processing. Coding ensures that transmission is spread over the entire frequency band allocated to a WHART network.

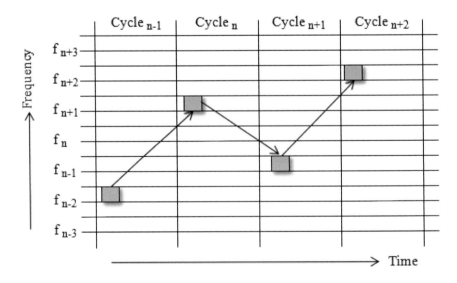

FIGURE 25.10   Channel hopping in WHART.

WHART-enabled field devices can correctly decode this coded informa-tion while others see the transmission as a white noise. As a consequence, multiple overlapping radio signals are received and properly decoded only by the WHART network devices.

25.7.3 LOW POWER TRANSMISSION

WHART devices are provided with low power compared with other wire-less applications like RFID readers, Bluetooth, etc. Thus, WHART devices are subjected to less interference from such applications. For process con-trol applications, IEEE 802.15.4 standard has been chosen for their rela-tively low power in which amplifiers can cover a distance up to 200 m. It acts as a router to pass the message along en route to the access point. For lesser distances, even lower power amplifiers are used for hopping from one device to the next one, thus reducing chances of interference even more. This limits RF pollution for users using the same spectrum.

25.7.4 BLACKLISTING AND CHANNEL ASSESSMENT

WHART can be configured to avoid channels which are relatively used more than others. This would result in lesser interference. But in reality, most networks are not used continuously and hence chances of such inter-ference is not very high.

Prior to transmission, a WHART device listens to the frequency chan-nel it intends to use. If another transmission is detected, the device will back off and would attempt transmission at a later time slot on a different frequency.

25.7.5 SPATIAL DIVERSITY

It involves placing wireless devices in such a manner that coexistence issues are minimized with nearby wireless devices. It requires careful planning during the installation phase and also during plant expansion.

One example of spatial diversity is antenna diversity in which the receiver receives the best signal from multiple antennas. Again directional antennas transmit in a specific direction which helps in mitigating RF interference.

25.8   TIME SYNCHRONIZED MESH PROTOCOL (TSMP)

Time synchronization is critical in WHART. Communications on WHART are precisely scheduled and timed. It is based on TDMA mechanism and employs a channel hopping scheme for system security. Communication schedules are structured by graph routes that create communication paths between one or more paths (hops). All network devices keep a track of time in terms of ASN which helps them to communicate with each other using the correct scheduled time slots. Gateway is the device which provides the time clock to the whole network.

Time is required for measurement data to reach the controller via different paths incorporating latency in the system. In WHART, each field device is time stamped with time accurate to 1 ms across the entire network. This frequent synchronism of the devices is not available with most of the other protocols.

25.9   SECURITY

WHART is an IEC-approved standard mesh networking technology and is used for comparatively secure and reliable message transmission in process control and automation environment. A field device can have one or more sensors which collect information about the process and send them to other field devices. This routing information, security keys, and the timing information are sent to the devices in a very secure manner. Data travel along the WHART network in the form of commands and confidentiality, integrity, and authenticity of the commands are ensured.

WHART neither enforces nor specifies any means to provide security and reliability in the wired part of the network unlike its wireless portion.

25.9.1   OSI Layer Based Security in HART and WHART

Figure 25.11 shows the security and reliability aspects as applied in HART and WHART at different layers of OSI protocol. In WHART, data from application layer down to transport layer is not provided with any cryptographic protective cover, although from network layer downwards, it is otherwise.

25.9.2   End-To-End Security

This is all about the security of data when it travels between source and destination devices. The network layer is responsible for providing this security.

A field device cannot create a session with another one. Thus, data from a source field device to a destination field device has to travel through a gateway. A session can be created between a field device and a gateway or between a field device and the network manager. Thus, data from a source field device is first encrypted with a unique symmetric session key and sent to the destination device via the gateway. The gateway decrypts data from

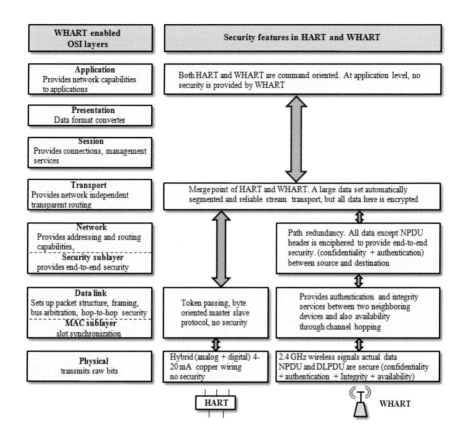

FIGURE 25.11 Security at different layers of HART and WHART based on OSI protocol. (Courtesy: S Raza. Secure Communication in Wireless HART and its Integration with Legacy HART, Swedish Institute of Computer Science, Technical Report T2010-01, ISSN: 1100-3154, SICS Networked Embedded Systems (NES) Laboratory, Kista, Sweden, p. 22, 2010.)

the source and encrypts it again with the session key of the destination device. The gateway finally sends it to the destination device.

25.9.3 NPDU

The network layer PDU (NPDU) is already shown in Figure 25.6 and consists of several fields which are shown as header field in Figure 25.12. The security sub layer of NPDU consists of three fields: a security control byte, a counter, and the MIC.

The NPDU Payload shown is the transport layer PDU. It is encrypted using an AES with a 128 bit key. The other fields present in the NPDU are responsible for routing of data. Network layer provides confidentiality, integrity, and authentication.

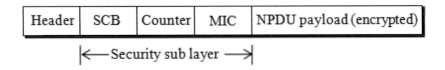

FIGURE 25.12   The WHART NPDU.

25.9.3.1   Security Control Byte

The security control byte (SCB) is part of the NPDU. It is used for defining the type of security employed. The upper nibble of SCB is reserved for future security requirements while the lower nibble defines the security type which are: session key, join key, and handheld key.

25.9.3.2   Message Integrity Code (MIC)

MIC provides source integrity (authenticity) and data integrity. Four byte strings are needed for MIC calculation which are: NPDU header, NPDU payload, AES key, and the nonce. AES key is used for encryption of NPDU payload.

Nonce is 13 byte in length and is used to defend reply attacks. The first byte is either all ones (for join response message only) or all zeroes. The nonce counter consists of the next 4 bytes. The last 8 bytes comprise the source address.

25.10   SECURITY THREATS

Security is required in both the wired and wireless portions of an industrial process control and automation system. As WHART shares the same ISM band along with other wireless communication systems like Wi-Fi, Bluetooth, ZigBee, etc., there always remains a security threat from these. Presence of unwanted signal may lead to security breach in the form of information leak, disruption in network traffic or a change in information at the receiving end. Security threat is also encountered at the point at which the wireless part of the network meets its wired counterpart. Figure 25.13 shows the different kinds of security threats and ways to combat them.

25.10.1   INTERFERENCE

It is an unforeseen and *unintentional* disruption to the intended communication signal which is meant for the receiver. Possibility of such an interruption overriding and dwarfing the original signal occurs when the interfering signal has the same frequency and modulation as that of the

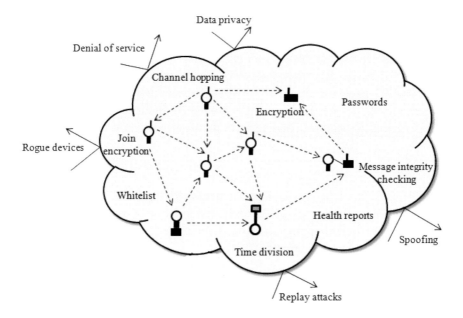

FIGURE 25.13 Different security threats and ways to combat them.

original signal. Any system operating with Bluetooth, ZigBee, and Wi-Fi may cause interference in the vicinity of a WHART system. It may be mitigated by using frequency hopping spread spectrum (FHSS), time diversity, and path diversity techniques.

25.10.2 Jamming

In this, an *intentional* disruption to the network signal is introduced in the form of some noise having the same frequency and modulation technique as used in WHART network. The effect of such jamming on the network can be severe and damaging to the network performance and it depends on the intruder's ability to inject noise signal to jam the network.

To overcome this, WHART uses FHSS technique which *blacklists* the channel(s) which is/are constant source(s) of noise. But such blacklisting of channels results in lesser number of available channels to carry WHART signals. Blacklisting of channel is done manually by the network administrator.

25.10.3 Sybil Attack

In this, an attacker may hold multiple identities in the form of a node or a software introduced in the system. Such sybil attack on the system may hamper the performance of the network badly. The attack can be taken care of effectively in the following manner.

Each device used in the network has a global unique ID, which is a combination of device type and device ID. These unique IDs are maintained by WHART gateway, while the network manager assigns individual nicknames to each device. They use these unique ID and nickname to establish sessions with the gateway and network manager, respectively.

25.10.4 COLLUSION

Collusion occurs when more than one device try to access the same channel at the same time. It can either be intentional or otherwise and can be reduced by channel hopping and time diversity technique.

It is detected by CRC and its effect on system performance is reduced by proper coordination and implementation of physical and data link layers.

25.10.5 TAMPERING

If an intruder has knowledge of network key and unencrypted DLPDU, the latter can very easily be tampered with, and a new MIC formed to make it appear authentic. The attack can be serious and damaging since the DLPDU is unencrypted and the packet can be sent to another destination. The attack becomes more damaging when the packet is sent back to the source. This can have adverse consequences on network performance.

25.10.6 SPOOFING

A device intending to join the network does so by advertisement time slots earmarked for the same. This is done with the help of the well-known network key. An attacker can spoof a new device intending to join the network by sending a fake advertisement. On receipt of the join request, it can simply reject the new device and debar it from joining the network. Proximity of the spoofing device to the device intending to join the network can permanently discard the latter from joining the network. If the spoofing device has access to the network key, it would result in more malicious attacks leading to network traffic blockage.

25.10.7 EXHAUSTION

An attacker can use a device to send messages to other devices in the network by using the well-known network key. This is possible when the attacking device has knowledge of network parameters and conforms to WHART protocol stack. Using network key, the MIC of the DLPDU can be calculated and the attacking device would then use the join key to encrypt and authenticate NPDU. Such fake messages, however, would be discarded by the network manager finally, but it would consume both network resources and time thereby seriously affecting network performance and block network traffic. Flooding a network with such repeated join attempts would lead to severe jamming of the network.

25.10.8   DOS

DOS or denial of service, can be effected in any one of the following ways:

- Sending fake time advertisements
- Flooding the network with fake join requests
- Jamming the network signal
- Replacing the unencrypted DLPDU and recalculating CRC

In such cases message integrity can be calculated by MIC, which can be found out with the help of AES in CCM. But verification of MIC involves a lot of network time which would eventually reject the unverified packet.

25.10.9   TRAFFIC ANALYSIS

It can be done on NPDU and DLPDU which are unencrypted. NPDU header fields involve source and destination addresses, security control byte, ANS snippet, nonce counter which are sent in clear. Again DLPDU header fields involve address specifier and DLPDU specifier which are also sent in clear. These fields in the two layers can easily be analyzed by the attacker. The attacker would then be able to know the device usage rate, peak usage period, join requests, etc. to launch attacks much more effectively to bring down network performance.

25.10.10   WORMHOLE

In this, the attacker creates a tunnel between two legitimate devices by connecting them through a stronger wired or wireless link. The wormhole attack is launched through a HART device which is connected to WHART via adapters. The tunnel is established between these two devices by using their maintenance ports, although WHART denies establishing such a tunnel by restricting network access in this mode. Such a tunnel can be created by wireless connection if the network and session keys are known to the attacker.

If WHART uses *Graph Routing* (it supports redundant paths), then the system is prone to wormhole attack. *Source Routing* uses a device-by-device route from source to destination which can repulse a wormhole attack. But such a routing scheme is unreliable since if any one interme-diate link fails, it would lead to a packet failing to reach the destination. Packet leaching method is used to ward off a wormhole attack.

25.10.11　SELECTIVE FORDWARDING ATTACK

It can occur only if there is a sybil attack on the network. Here, the com-promised node selectively drops packets, instead of forwarding all. A black hole is created when no packet is forwarded. Normally the attacker selectively drops packets so that it appears to be a genuine and cannot be retrieved by the recovering mechanism.

25.10.12　DESYNCHRONIZATION

A WHART network is based on a strict time synchronism principle. A timer module maintains this time synchronism of 10 ms. MAC sublayer takes care of this. When a node receives an ACK, it adjusts its clock accordingly. The source of timing can be a sender and if the sender is com-promised, it can disrupt the timing between the two nodes. Thus, the two nodes would try to adjust their time synchronism and precious resources are simply wasted.

25.11　REDUNDANCY

Data and information flow at all levels in a process control system is a must for proper control. Redundancy is used at places where process data is operationally very critical or the failure of a device leads to considerable loss of data. Adding redundancy to a system increases probability of data availability even if a device fails or else normal data communication path is blocked due to some reason whatsoever. WHART provides redundancy at the following levels:

- In the wireless sensor network
- At the network access points
- At the gateway, network manager, or security manager

25.11.1　REDUNDANCY IN WSN

Redundancy in wireless sensor network (WSN) can be achieved by any one of the following manner:

- **Spatial diversity:** using multiple paths between network components
- **Frequency diversity:** using frequency hopping technique
- **Time diversity:** accessing network at different times

Device redundancy can be achieved by providing multiple paths to each device for data to reach the gateway.

25.11.2 REDUNDANCY AT NETWORK ACCESS POINTS

Multiple access points allow additional path diversity and hence redundancy to the system. A network access point provides a communication path between a device and a gateway via the access point. An access point can be thought of as a network device having a higher bandwidth. It provides path diversity as well as redundant communication paths for the higher level gateway and network manager.

Theoretically a network can have any number of access points, but finally a tradeoff is made between cost, overall network bandwidth, and redundancy requirements.

25.11.3 REDUNDANCY AT GATEWAY, NETWORK MANAGER AND SECURITY MANAGER

Redundancy can be enforced at gateway, network manager, and security manager by clubbing them together in one physical gateway device and repeating this block all over again. The same is shown in Figure 25.14. Redundancy decision is taken by the designer of the network who has knowledge about the most critical components.

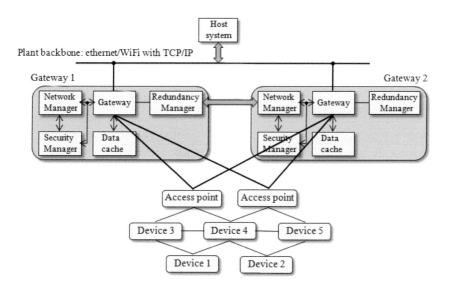

FIGURE 25.14 Redundancy at gateway, network manager, and security manager.

In the figure, one gateway acts as primary and the other takes over once the running gateway fails. The two gateways communicate with each other via a redundancy manager which keeps the system synchronized.

25.12   SECURITY KEYS IN WHART

A comprehensive key management scheme is not there for both wireless and wired portions of WHART. WHART standard specifies the need to have a security manager, but its design and functionalities are not explicitly mentioned.

Keys used in WHART have the following features:

* Utilizes AES-128 block ciphers, unsigned with symmetric keys
* Supports multiple key architecture
* Separate join key per device
* Data link PDUs authenticated by network key
* End-to-end communication authenticated by session key

WHART standard specifies different keys and their functions, but does not mention the manner of generating, storing, revoking, renewing, or distribution of the keys. The standard specifies that the network manager is responsible for sending keys to the devices via the gateway while the security manager is used for generation, storage, and vetting of keys. WHART standard does not explicitly specify the manner of distribution of keys, the interface between the network manager and security manager. Different keys used in WHART are: (a) join key, (b) session key, (c) network key, (d) handheld key, and (e) well-known key.

Security provided by WHART is transparent to application layer. Each join key has to be distributed to every device prior to network initialization. Security keys are renewed and rotated as per process requirements.

25.12.1   JOIN KEY

Each device in WHART network has its individual join key to join the network. It acts as a password to the network. The security administrator manually distributes this key to each device. The handheld device is normally used to enter the join key to the device using the maintenance port of the device. When the join key is being added to the device initially or changed later, the device must remain isolated from the network. The join key is a *write-only* key, like all other keys in WHART.

Network manager can change a join key later. It also authenticates the new device using the join key and on successful authentication, writes the network key and session key into the device.

25.12.2   SESSION KEY

Each device has four session keys which are allocated to them after successful joining, restart, or after power on. The four session keys are: (a)
unicast session key between the gateway and the device, (b) unicast session
key between the network manager and the device, (c) broadcast session key
from the gateway to all the devices, and (d) broadcast session key from the
network manager to all the devices.

Two field devices cannot communicate directly with each other, but do
so via the gateway. The first (source) field device sends the message to the
gateway via the unicast gateway session key of the source device. The gateway then transfers it to the second (destination) device using the unicast
gateway session key of the destination device.

25.12.3   NETWORK KEY

There is only one network key which is shared by all the devices. It provides protection to DLPDU against outside attack. It is used to calculate
the keyed MIC which protects and secures DLPDU. Two communicating
devices authenticate each other on the basis of this MIC.

25.12.4   HANDHELD KEY

It is provided by the network manager using the join key, when requested
by the handheld device. It is used for peer-to-peer connection between a
handheld device and a device belonging to the network. This mode is used
for device maintenance by using the device's maintenance port. Such a
connection does not involve a gateway. The handheld key is used to secure
the NPDU.

25.12.5   WELL KNOWN KEY

All messages in a WHART network must be encrypted. During the join
process, a device has only the join key which protects the NPDU but does
not possess the network key to protect DLPDU. A network key, called
the well-known key, is used to calculate the MIC for the join request/
response messages. The well-known key is also used for sending join
advertisements.

25.13   KEY MANAGEMENT

Automatic key management is a vital area without which network
security would be jeopardized. Manual key adjustments would lead to

occasional system operation failure. WHART employs network manager for various key distributions, i.e., it acts as a *key distribution center*, while security manager is employed for key management purposes, but the key management issue is not very comprehensive in WHART network. Key management involves the following: (a) key generation, (b) key storage, (c) key distribution, (d) key renewal, (e) key revocation, and (e) key vetting.

The keys used are symmetric in nature, i.e., it does not use asymmetric cryptography.

25.13.1 KEY GENERATION

Key generation is the responsibility of the security manager. WHART standard does not specify explicitly ways for key generation but specifies the keys needed to secure the channels.

25.13.2 KEY STORAGE

Network security in a WHART system is dependent on keys. Storage of keys in a secure manner is of paramount importance, although WHART does not provide means for key storage, like key generation.

25.13.3 KEY DISTRIBUTION

In a WHART system, distribution of keys to the devices is the responsibility of the network manager. How the network manager requests the keys from the security manager remains a grey area and not specified fully.

The WHART standard provides information regarding how the join key is provisioned in a device and provided to the network manager.

25.13.4 KEY RENEWAL

Any key in WHART is breakable using the *Brute-Force Attack*, provided there are no time or resource constraints. Key renewal or key changing is very important so that outside attacks can be repulsed. Thus, all keys should frequently and automatically be changed. WHART has mechanism to do the above, i.e., to alter or renew the keys.

While the network key can be changed by using a secure broadcast session, other keys can be changed by using unicast session between the network manager and the devices. Security of join key and session keys are interdependent. Hence, lack of security of one would affect the security of the other.

25.13.5 KEY REVOCATION

Key revocation means deactivation of a key when its corresponding device has left the network. If the device leaves the network permanently, the network key should also be changed. A device, initially, has only one secret key which is the join key. When the device joins the network with its help, the device gets the network and session key details.

In a WHART network, each key belongs to a particular device. Thus, if a device is captured, its authenticity is lost and the device should *self-destruct* for security reasons. Else the key must be automatically deleted from the network.

25.13.6 KEY VETTING

Authentication or verification of a key is called key vetting. It is mostly used in public key infrastructure (PKI). No specific command is there in WHART for key vetting purpose.

25.14 WHART NETWORK FORMATION

Network formation in WHART has to go through several stages before the same becomes ready for industrial use. Network manager sets the network ID and password, followed by join process and session security. At the shop floor, password and network ID of the device are entered. The device is then configured as required and the update rate is specified

After this initial setup, the network manager sends an *advertisement* to the concerned device and the device responds with a join request.

In the third and final stage, the network manager authorizes the device with particular keys, schedules, and routing. On the basis of this information, the device can publish data.

25.15 HART AND WHART—A COMPARISON

A comparison between HART and WHART is shown in Table 25.1. WHART is introduced because of increasing applications of wireless technology in industrial automation and control systems.

25.16 HART AND WHART—INTEGRATION

Legacy HART protocol is there around for some 25 years while its latter counterpart WHART is about a decade old. There is a huge industrial base for the HART in the industrial control and automation sector. Thus, a need is felt that existing HART network in a plant are integrated with

TABLE 25.1

Comparison between HART and WHART

HART	WHART
Not a secure protocol	A secure protocol
Legacy HART belongs to HART 6 and below	WHART belongs to HART 7 and above
Based on OSI 7 layer protocol, but loosely coupled	Based on OSI 7 layer protocol, but strongly coupled
Transport and higher layers (up to application layer) identical to WHART	Transport and higher layers (up to application layer) identical to HART
Application layer consists of command number, byte count, and data	Application layer consists of command number, byte count, and data
A 16-bit command number	A 16-bit command number
Does not explicitly specify network or higher layer except application layer	Uses network layer for wireless routing and end-to-end security
Based on token passing data link layer	Based on TDMA data link layer
Physical layer based on 4-20 mA analog wiring	Physical layer based on IEEE 802.15.4-2006

WHART being introduced in the same for either extension or modification purposes. Integration between HART and WHART is possible due to the following:

- WHART devices and networks are backward compatible with existing HART.
- Since persons are used to HART, it is easy for them to embrace WHART.

Integration between HART and WHART was undertaken by HCF.

Integration between HART and WHART networks is done by using a gateway. They have different protocols and a gateway acts as a protocol converter to integrate the two. A WHART gateway can be used in a PC card, as part of an I/O subsystem or as a standalone device.

26 ISA100.11a

26.1 INTRODUCTION

ISA100.11a is a wireless standard developed by International Society of Automation (ISA). It was endorsed as an ISA standard in September, 2009. The standard provides reliable and secure wireless communication in the field of industrial automation for noncritical monitoring and control applications. ISA100.11a can integrate with HART, Foundation Fieldbus, PROFIBUS, and others through device adapters, network protocol pass through tunneling, by mapping using interface objects, etc.

26.2 SCOPE OF ISA100

ISA100 is one of the three major competing standards in wireless sensing and communication area and formed in 2006. It is managed by ISA. Automation Standards Compliance Institute (ASCI) was formed by ISA for certification, conformance testing, and compliance assessment of the activities being carried out in ISA's automation field. It acts as a bridge between testing of the ISA standards and their effective implementation. Again Wireless Compliance Institute (WCI) operating under the fold of ISA 100 is responsible for compliance certification of ISA100. Activities of WCI include:

- Interoperability assurance via testing, standards and conformance testing
- Conducts independent testing and certification of field devices and systems to ISA100 family of standards
- Users and suppliers updated with technical support, text materials, and tools
- Certifies devices and systems with regard to their meeting a common set of specifications and standards
- Sensor node network connectivity
- Host node network connectivity
- Multi-vendor device interoperability among field devices
- Multi-vendor device interoperability between routers and field devices
- Multi-vendor device interoperability between gateways and devices
- Sensor data flow
- Network health metrics, metric collection, and presentation
- Field device parameterization/configuration

26.3   ISA100 WORKING GROUP

The broad objective of ISA100 is to establish standards, prepare technical reports, recommend allowed practices, provides information for correct implementation of wireless communications in process industries. A number of working groups were formed to address the above tasks.

26.4   FEATURES

ISA 100.11a is a highly reliable, network scalable wireless technology based on IEEE 802.15.4 standard physical layer and supports star, mesh, and star-mesh networks. It supports redundancy and interoperability and can blacklist channels that interfere with data transmission. Real-time data transfer is based on time division multiple access (TDMA) method and has a variable slot time. It uses channels 11–25 as defined by IEEE 802.15.4 with optional channel 26. Typical number of field devices may be 50–100 which may or may not have router capability. A combination of FHSS and direct sequence spread spectrum (DSSS) is used for the purpose of modulation. Channel hopping takes place on a packet-by-packet basis that supports slotted, slow, or a combination of these two. It supports five pre-programmed hopping patterns. It does not have presentation and session layers of the open systems interconnection (OSI) model, but its application layer has two sublayers and three data link layers. It checks if a channel is *free*, before sending data called clear channel assessment (CCA). Activities are fully time synchronized maintained by an internal clock. Addressing modes supported are 128-bit long address (IPv6) and 16-bit short address. DLL supports either graph or source routing. Devices consume very low power and the battery lasts around 2–5 years.

26.5   SENSOR CLASSES

There are six different sensor classes defined by ISA100 committee. These are class 0 to class 5 with class 0 the most sensitive one. It is normally not recommended to transport data over this class. Classes 0–3 correspond to *control* while classes 4 and 5 are used for *monitoring*. Importance of message timeliness increases from class 5 towards class 0.

26.6   SYSTEM CONFIGURATION

Figure 26.1 shows a typical system configuration based on ISA100.11a. It consists of field devices, backbone routers, a backbone network, gateway, system/security manager, control system, and applications.

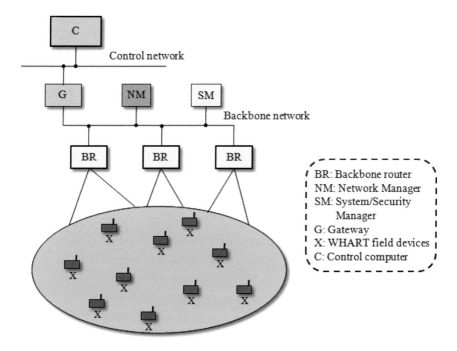

FIGURE 26.1   System configuration based on ISA 100.11a.

It is seen from the figure that redundant paths exist between backbone routers and field devices on the one hand and between controller and field devices via backbone routers on the other. The gateway acts as an interface between the wireless network and the control applications such as a distributed control system (DCS). System security is managed by security manager at every level of the network whereas system manager is responsible for controlling and managing the activities of the total network.

26.7   CONVERGENCE BETWEEN ISA100.11A AND WHART

In order to achieve convergence between ISA100.11a and wireless highway addressable remote transmission (WHART), ISA100 formed a convergence subcommittee, designated as ISA100.12. This subcommittee solicited suggestions, requirements, and other inputs from the end users of ISA100.11a and WHART through a Convergence User Requirements Team or CURT. User requirements obtained by CURT were then directed to ISA100 and also to NAMUR for implementation of the recommendations. The subcommittee developed a request for proposals (RFP) which would be responsible for ultimate convergence of these process control standards. RFP was passed in September, 2010.

26.8   NAMUR PROPOSAL

It is an international association of process and automation industry end users—mostly European. NAMUR collects inputs from the users in the form of drawbacks of existing systems, possible ways out to remove them, suggestions for improvements that may be implemented in future. These inputs are assorted, works are published by NAMUR and forwarded to standardization authorities for possible standardization after incorporating suggestions or alterations as suggested by authorities.

Three emerging standards in the wireless arena in the process industry sector are: IEC/PAS 62591 (WHART), WIA-PA (IEC/PAS 62601), and ISA100.11a. Convergence of these three standards is of prime importance which would enable the end users to get over the interoperability problem.

Convergence is of utmost importance so that end users are least affected with regard to the standard that is being used by them. It is suggested that a single standard be developed in place of multiple standards that exist at the moment.

26.9   ARCHITECTURE

The detailed architecture of ISA100.11a is shown in Figure 26.2. ISA100.11a is the first wireless protocol from ISA100 family. It is an evolving standard in the process automation and control field with reliable and secure wireless communication. Different types of devices have been shown, along with subnets, backbone routers, backbone network, gateway, system manager, security manager, and control network.

Data from a device can reach routers via different paths—it depends on availability and loading pattern of the paths. Security manager is responsible for all security-related issues for secure communication over the network. It manages the security keys and coordinates with the system manager for proper system operation—both internally and externally. Information related to security between different devices and the security manager is carried out by proxy security management object (PSMO).

The system manager, which acts as a system administrator, is responsible for system management, device management, resource management, time-related issues, etc. This is implemented by the system manager with the help of a system management application process (SMAP).

The function of gateway is almost identical to the functioning of the gateway of WHART. The gateway directly interfaces the host-level applications to the wireless field devices or to wired devices via adapters. The gateway is implemented with a protocol translator and a gateway service access point (GSAP) at the application layer based on existing protocol suite.

The different devices and their roles and responsibilities are tabulated and shown in Table 26.1.

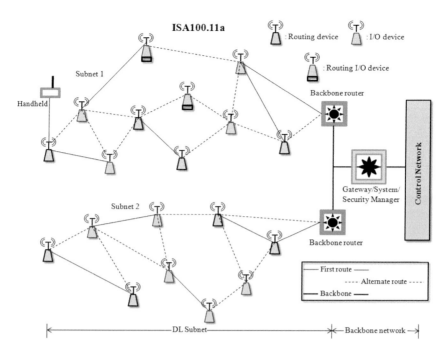

FIGURE 26.2 Detailed architecture of ISA100.11a. (Courtesy: G. Wang. Comparison and Evaluation of Industrial Wirelessv Sensor Network Standards: ISA100.11a and WirelessHART. Master of Science |Thesis, Communication Engineering, Chalmers University of Technology, Sweden, 2011)

TABLE 26.1
Devices and Their Respective Roles/Responsibilities

Device type	Roles/Responsibilities
Input/output	A source or sink of data
Router	Routes a message through the nodes in the same subnet
Backbone router	Routes data from one subnet to another
System manager	It is central to all operations in the network. It manages all network activities through pre-configured software
Security manager	Enables, controls, and supervises all security issues of all devices present in the network
Gateway	Provides an interface between wireless and plant network
Provisioning device	Provisions devices with configurations to ensure proper system operation
System time source	Maintains the master time source in the network

Source: ISA, The Technology behind the ISA100.11a Standard-An Exploration. ISA100 Wireless Compliance Institute, 2010.

26.9.1 Differences With WHART

Table 26.2 shows the architectural and OSI layer based differences between ISA100.11a and WHART.

26.9.2 Routing Ability of Devices

In ISA100.11a, field devices can be implemented with or without router capability. Devices can be assigned flexible roles—i.e., multiple role profiles having different functionalities. A device with router capability can

TABLE 26.2

Architecture and OSI Level Based Differences between ISA100.11a and WHART

	ISA100.11a	WHART
Architecture level differences	Mesh, star, or star-mesh network	Only mesh network
	Backbone devices	Network access points
	Provisioning device with DPSO	Handheld used for commissioning
	QoS with secure communication subnets within the total network. The architecture thus becomes scalable.	Peer-to-peer communication with potential security risks
	Devices have routing capability	
Digital Link Layer	Five preprogrammed hopping schemes	One channel hopping scheme
	Configurable length of time slot	Fixed time slot
	Active and passive neighbor discovery	Passive neighbor discovery
Network Layer	Three header specifications	One header specification
	Based on IPv6 addressing	Based on HART addressing
	Fragmentation and assembly	End-to-end security
	Compatible with 6LoWPAN	
Transport Layer	Connectionless service (UDP), unreliable	TCP-like reliable communication
	Compatible with 6LoWPAN	
	End-to-end session security	
Application Layer	Object oriented	Command oriented
	Support legacy protocol tunneling	Support HART protocol
	Standard management objects, dependent	Predefined data types
	Independent objects, ASL services	

Source: G. Wang. Comparison and Evaluation of Industrial Wireless Sensor Network Standards: ISA100.11a and WirelessHART. Master of Science |Thesis, Communication Engineering, Chalmers University of Technology

improve the working of network mesh by reducing latency. Router role of a device can be switched off by the system manager to increase energy savings. Again, in harsh environment, devices with router capability can improve mesh networking ignoring power saving aspects.

26.9.3   SUBNET

In ISA100.11a, the whole network is divided into smaller ones. Each such network is called sub network or simply a subnet. An ISA100.11a can have multiple subnets with each subnet having a maximum of 2^{16} devices. Thus, within a specific subnet, a local unique 16-bit address is assigned to every device by the system manager. Data routing takes place both at the local or subnet level using the 16-bit short address and a 128-bit long address at the network level. Thus, scalability in ISA100.11a by having subnets gives it a more integrated approach to the overall message transmission in the network.

26.9.4   PROVISIONING DEVICE

A device intending to join a network has first to be provisioned by a provisioning device. The credentials of the new device are checked by the provisioning device when the new device requests to join via the join process. The join process is carried out by a device provisioning service object (DPSO). Asymmetric key based OTA (over the air) provisioning and open symmetric join key provisioning methods are undertaken for a new device to join the network.

26.9.5   BACKBONE ROUTERS

It acts as an interface between field network and control network. Backbone routers help in minimizing system latency, increase bandwidth, and improve quality of service (QoS). It encapsulates network layer data and sends the same through the protocol stack of backbone to the destination. Use of backbone routers help in lesser traffic congestion and thus reduce latency. ISA100.11a has already established configuration between backbone routers and other devices in the network for proper integration.

26.9.6   DEVICE MANAGEMENT DATA FLOW

Every device in the network is managed by a device management application process (DMAP) which also oversees communication aspects as well. Since application layer of ISA100.11a is object oriented, DMAP has a series of objects which support device management operations by proper

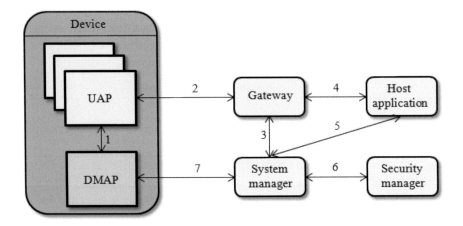

FIGURE 26.3 System management data flow diagram. (Courtesy: G. Wang. Comparison and Evaluation of Industrial Wireless Sensor Network Standards: ISA100.11a and WirelessHART. Master of Science IThesis, Communication Engineering, Chalmers University of Technology, Sweden, 2011)

configurations and invoking methods via management service access points at the respective layers. The network's system manager can also perform device configuration remotely via ASL (application sublayer services).

Management data flows through protocol layers via respective data service access points (DSAPs) while on the other side, DMSAP manages activities up to the level of upper data link layer.

26.9.7 SYSTEM MANAGEMENT ARCHITECTURE

The system management architecture of ISA100.11a is shown in Figure 26.3. Each device belonging to the network can be thought of as comprising of two blocks—DMAP and user application process (UAP). The DMAP is responsible for communication aspects and is administered by system manager via ASL services. System manager operates in conjunction with security manager.

UAPs are resident in the devices themselves. They are controlled by host applications via the gateway. DMAP and UAP communicate between themselves for overall device management.

26.9.8 SYSTEM MANAGEMENT APPLICATION PROCESS

System manager is in charge of overall system management of the wireless network. It acts as the system administrator and is responsible for system management, device management, communication between devices and the network, timing control, etc.

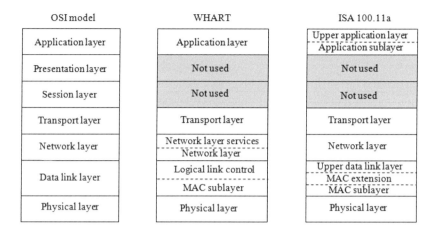

FIGURE 26.4 Comparison of protocol stack of ISA100.11a and WHART.

26.10 COMPARISON BETWEEN ISA 100.11A AND WHART PROTOCOL STACKS

Figure 26.4 shows the differences in the protocol stack of ISA100.11a and WHART with reference to the seven layer OSI model. Both ISA100.11a and WHART use the simplified version of the OSI model and they neither use presentation layer and session layer. Application layer of ISA100.11a is divided into upper application layer and application sublayer while the data link layer is divided into upper data link layer, media access control (MAC) extension and MAC sublayer.

Different standards are adopted at different layers of ISA100.11a. These standards strictly adhere to industry accepted norms. It is shown in Figure 20.5 along with OSI layer.

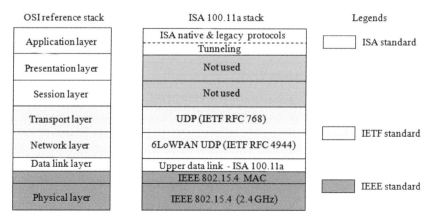

FIGURE 26.5 Different standards in ISA100.11a stack. (Courtesy: ISA, The Technology behind the ISA100.11a Standard-An Exploration. ISA100 Wireless Compliance Institute, 2010.)

26.11   PHYSICAL LAYER

Physical layer of ISA100.11a is based on IEEE STD 802.15.4. It acts as the interface to the physical medium through which actual message transmission occurs. It transmits and receives raw data packets, performs CCA, provides control mechanism, and detects RF energy. Direct sequence spread spectrum along with frequency hopping spread spectrum techniques are used for modulation purpose. A bit rate of 250 kb/s with offset—quadrature phase shift keying (O-QPSK) is allowed with a power limit of 10 mW. It uses channels 11—25 with optional channel 26 in the 2.4 GHz ISM band. Each channel uses a bandwidth of 2 MHz and channel separation of 5 MHz between any two successive channels.

26.12   DATA LINK LAYER

Data Link Layer of ISA100.11a is divided into three sublayers: upper data link layer, MAC extension, and MAC sublayer. DLL's main responsibilities are: access and synchronization, handling acknowledgement frames, and a reliable link between two peer-to-peer radio entities.

The MAC sublayer is a subset of IEEE 802.15.4 MAC. Its main responsibility is sending and receiving data frames. The MAC extension includes some additional features which are not supported by IEEE 802.15.4 MAC. This includes changes in carrier sense multiple access with collision avoidance (CSMA-CA) mechanism by including additional frequency, time, and spatial diversity.

The upper data link layer is responsible for routing within the DL subnet and takes care of link and mesh aspects of the subnet-level network. The system manager is responsible for DL subnet operations. The DL subnet ends at the backbone router. Networking beyond this is handled by the network Layer.

26.12.1   PROTOCOL DATA UNIT

The protocol data unit of data link layer (DPDU) is shown in Figure 26.6. It is more complex than WHART DPDU, but offers more flexibility and better performance. DPDU has a sub-field called data link layer header (DHR) which takes care of different functions like routing, congestion control, optional solicitation message, security aspect, mesh, and link.

26.12.2   COEXISTENCE ISSUES IN DLL

DLL operates in the 2.4 GHz range of industrial, Scientific and Medical (ISM) band and has to coexist with others like Bluetooth, ZigBee, WiMAX,

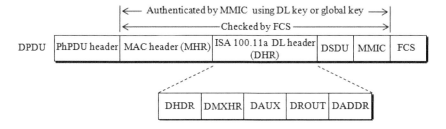

FIGURE 26.6 DPDU of ISA100.11a. (Courtesy: G. Wang. Comparison and Evaluation of Industrial Wirelessv Sensor Network Standards: ISA100.11a and WirelessHART. Master of Science |Thesis, Communication Engineering, Chalmers University of Technology, Sweden, 2011)

etc. of IEEE 802.11 b/g. For proper communication in ISA100.11a, several methods like channel blacklisting (to avoid channels which are very much in use), TDMA technology (to minimize possibility of collisions), channel hopping (to reduce multipath fading), data authentication and integrity (to maintain data confidentiality), etc. are followed. ISA100.11a adopts several mechanisms to overcome the coexistence issues so that highly reliable communication takes place in presence of other communications taking place at the same time in the same ISM band. Coexistence mechanisms utilized for increased data reliability include time diversity (TDMA), collision avoidance technique, frequency diversity and frequency diversity with ARQ (channel hopping), spectrum management by channel blacklisting and adaptive hopping, etc.

26.12.2.1 TDMA

WirelessHART TDMA technology uses a fixed timeslot of 10 ms for network-wide time synchronization. ISA100.11a, on the other hand, uses a configurable timeslot with duration varying between 10 and 12 ms on a per superframe basis. A variable timeslot gives rise to optimized coexistence with shorter timeslots using tight synchronized communications while the longer ones can be used for accommodating serial ACKs from multiple destinations, CSMA's performed at the beginning of a timeslot etc.

26.12.2.1.1 Slotted

The slotted hopping scheme uses different channels in each successive timeslot. A DPDU with its ACKs/NAKs is sent in a timeslot. Messages, which need to be sent in a synchronized manner, use slotted hopping. It is also used where energy limitation is of prime importance. All channels, i.e., from 11 to 26 take part in slotted hopping.

In this case a device is forced to wait until its scheduled turn comes again leading to latency.

26.12.2.1.2 Slow

Slow hopping scheme, as the name suggests, hops from one channel to another in a slow manner. In this scheme, a particular channel is used for several consecutive timeslots. Thus, a particular channel's usage may range from 100 to 400 ms. It utilizes channel numbers 15, 20 and 25 normally. Channel hopping is slower than slotted hopping. The hopping period is configurable.

Devices having imprecise time settings not requiring tight time synchronization or receivers having considerable energy utilize this scheme for message transmission. This scheme is better suited for event-based traffic, where occurrence of an event may demand immediate transmission of a data packet. Devices earmarked as receivers are to be kept on for considerable length of time to listen to incoming traffic, thereby increasing power consumption of the receivers.

26.12.2.1.3 Hybrid

Hybrid channel hopping is a combination of slotted and slow hopping schemes. Thus, it enjoys the advantages of both—synchronous and regular message transfer of slotted hopping and less predictable messages like alarms, etc. of slow hopping.

26.12.2.1.4 Default

ISA199.11a employs five preprogrammed hopping patterns supported by all the devices in the network. The channels used in the different patterns are prefixed.

System manager *configures* the network to be used for any one of the above five patterns. Patterns 3 and 4 are meant for slow hopping while pattern 5 is used for enabling coexistence with WHART.

26.12.2.2 Collision Avoidance

It uses media contention mechanism using both TDMA and CSMA-CA which considerably increases data transmission reliability. The transmitter employs a clear channel assessment mechanism to first ensure that a channel is free for transmission. Otherwise, the transmitter backs off, disallowing transmission. A retransmission is attempted at after a random amount of time.

26.12.2.3 Frequency Diversity

Hopping is undertaken to avoid channels of heavy usage to avoid interference. Thus, it provides immunity against interference from neighboring channels and also robustness to mitigate multipath interference effects. The hopping sequence determines the channel that the transmitter is to utilize after completion of the current transmission. ISA100.11a employs several channel hopping schemes like slotted hopping, slow hopping, hybrid hopping, and adaptive hopping to address different communication needs.

In a subnet, devices use different offsets into the hopping sequence. This results in interleaved hopping. Again, automatic repeat request (ARQ) is enforced to ensure that unacknowledged data packets are retransmitted on a different channel utilizing a different frequency. Thus, frequency hopping, when employed with ARQ, increases reliability of data packet delivery considerably.

26.12.2.4   Spectrum Management

Data delivery in ISA100.11a is vastly improved by properly ensuring a channel's availability prior to data transmission. Each device in ISA100.11a maintains a statistical record of each channel's availability by way of CCA backoff counts, number of attempts made for transmission over a channel and number of NACKs received. The system manager takes the final call from the statistical record to blacklist a channel(s) for a particular length of time. Channels which are free from such interferences are used for data transmission. Such spectrum management is very effective in a noisy environment.

Adaptive hopping is a step ahead in channel blacklisting method where a particular device takes the decision locally by bypassing a channel based on information collected by the device about the channel. The device in this case can adapt itself so as to alter the hopping sequence which ultimately increases data transmission reliability.

26.12.3   ROUTING IN DLL

Routing in ISA100.11a is performed at the data link level with the help of graph routing or source routing which are configured by system manager. All devices in the network support the two routings at the DL level. A message, on reaching a backbone router, will be provided with a network layer header so that the message can move along across the network.

Routing from a device to the backbone router is based on device ID, destination address or else the default route. Again, a forwarding limit field in the DL header puts the maximum number of times a message can hop. As the message hops from one node to another, this number decreases. If it reaches zero, the corresponding DPDU is cancelled or discarded.

Depending on the neighborhood discovery, system manager always optimizes the route that a message would take to reach the backbone router.

26.12.4   NEIGHBORHOOD DISCOVERY

Neighborhood discovery helps in joining a new device into joining a network and maintain and upgrade the mesh network operation as devices join and leave the network. Its job are mainly:

- Designated advertisement routers advertise network information. The same is received by the devices desiring to join the network. The new device thus joins the network with the information it receives from the advertising router.
- A device, after it joins the network, sends/receives advertisements to/from neighbors. This helps the system manager to update itself so that network reliability and efficiency is improved. It is an ongoing process as devices continue to join/leave the network.

For effective neighborhood discovery, two schemes are followed in ISA100.11a. These are passive listening scheme and active scanning scheme. The former is identical to WHART. Active scanning comprises scanning interrogators and scanning hosts. The former seeks for advertisements from advertising routers. They also transmit advertisement solicitations to routers. Routers, which are called advertisement hosts, advertise the relevant information meant for the interrogators including subnet information helping the devices to join the network. The process of neighborhood discovery is scheduled and configured by the system manager.

Energy consideration is a very important aspect in neighborhood discovery scheme. Scanning hosts in active scanning scheme must have enough energy in their receivers to remain active during initial network formation and also during sensitive periods. Active scanning scheme has lesser latency compared to passive scanning scheme.

26.12.5   DLL Characteristics

Some of the characteristics associated with DLL are: configurable time slots of 10—12 ms—it thus can accommodate devices with different time requirements, supports unicast, duocast, and broadcast, neighborhood discovery involves active, and passive scanning, five predefined hopping patterns, routing in the network done via several subnet levels, etc. DL-level activities are maintained by system manager via DL management object (DLMO).

26.13   NETWORK LAYER

Some of the characteristics associated with the network layer are:

- It is influenced by Internet Engineering Task Force (IETF) 6LoWPAN specification
- 16-bit short address for DL subnets and 128-bit long address for backbone routers (IPv6).
- Fragmentation and reassembly of data packets

- Mesh-to-mesh routing
- Address translational mechanism

26.13.1 FUNCTIONALITY

Network layer takes care of addressing and routing at mesh level and backbone router level. At the backbone level, addressing and routing are done by IPv6.

Network level is responsible for correct address information—a 16-bit short address for DL subnets and a 128-bit long address for communication at the backbone level. Address conversion from DL subnet to backbone is done by backbone router. This address conversion is done by an *address translation table* maintained at the network level. Each device in the network maintains and updates their own address translation table with the help of system manager. Each entry in this table has a DL subnet address and its corresponding backbone address. Fragmentation and assembling of data packets are also carried out at this level with packet lengths more than that allowed at the data link layer level.

26.13.2 HEADER FORMATS

ISA199.11a supports both subnet level routing (mesh level) and backbone level routing. Communication can take place at intra subnet level or between two subnets taking the help of backbone routers in the process. For the former, a 16-bit address is all that is needed for data transfer requiring much less energy and bandwidth in the process. A 128-bit address needed for data transfer across two subnets require extra power and bandwidth.

Three different header formats are employed to address such diverse addressing needs. Depending on the level of service and routings required, designs can be optimized for both low power consumption and bandwidth. The three header formats are basic, contract enabled, and full IPv6.

26.13.3 DATA FLOW BETWEEN TWO SUBNETS

ISA100.11a may have a number of subnets—it is dependent on the physical placement of the devices and also on the total number of devices in the network. Data flow between two devices belonging to the same subnet takes place via the 16-bit DL address. Whenever data has to move from one subnet to another, it is done by a 128-bit long address by the backbone router for communication via the backbone. This is shown in Figure 26.7. The receiving end router converts the 128-bit long address with the 16-bit DL address of the device. Fragmentation and reassembling are done at the point where backbone level routing switches over to DL level routing and vice versa, respectively.

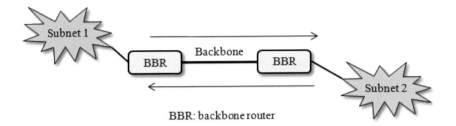

BBR: backbone router

FIGURE 26.7 Data flow between two subnets in a network. (Courtesy: G. Wang. Comparison and Evaluation of Industrial Wirelessv Sensor Network Standards: ISA100.11a and WirelessHART. Master of Science |Thesis, Communication Engineering, Chalmers University of Technology, Sweden, 2011)

26.14   TRANSPORT LAYER

It transfers data between end systems or host to host but not on the routers. It is responsible for end-to-end error recovery. It uses user datagram protocol (UDP) for connectionless service over IPv6 with optional compression as defined in IETF 6LoWPAN specifications. It offers better data integrity, encryption, and data authentication. Some of the jobs performed by transport layer are: reliable but unacknowledged service, flow control, segmentation and reassembly, basic and enhanced security service which is optional using session key.

26.14.1   PROTOCOL DATA UNIT

The transport layer protocol data unit (TPDU) is shown in Figure 26.8 and consists of four fields. UDP at this layer is normally not compressed but optionally can be compressed when the TPDU is passed on to the network level. It should be noted that the uncompressed UDP header requires a mandatory checksum in the form of IPv6.

The security header contains several security aspects in the form of different security levels, key-index mentioning the current session key and timing information. The third field is the application layer payload that is encrypted by using the session key. The fourth field, TMIC, is generated using the session key used for data authentication

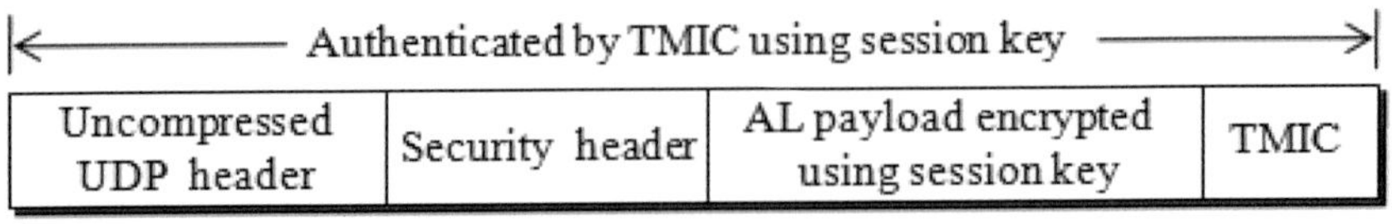

←——————— Authenticated by TMIC using session key ———————→			
Uncompressed UDP header	Security header	AL payload encrypted using session key	TMIC

FIGURE 26.8   The TPDU of ISA100.11a.

26.14.2 SECURITY

AES128_CCM in conjunction with UDP provide security in the transport layer for various types of communication services and for various levels of security requirements. Various options are enforced for the above such as:

- Transport layer MIC (TMIC) is computed with the help of session key for end-to-end data integrity protection but the UDP checksum is not enforced for full header compression as required in 6LoWPAN.
- During join process, the UDP checksum is incorporated for end-to-end data integrity but TMIC cannot be enforced due to lack of availability of session key.

Data encryption, data authentication, and the level of security to be enforced are all dependent on the security policy that is already in place during contract.

Transport layer provides the required security against any replay attack by using a time stamp in the nonce which indicates the time of formation of the data packet. If the data packet was created earlier than the configurable time limit, the receiver simply rejects it.

26.14.3 SESSION AND CONTRACT

Communication between two devices is executed with the help of security manager which is assigned a contract for the same. Several contracts may be in place and exist in the system manager to support different communication needs. Similarly, a session must be there between a source and the destination (end-to-end devices). The session key is granted permission by the system manager for data flow to take place. The master key in each device is used for any new session key or updating the same.

26.15 APPLICATION LAYER

Application layer provides necessary communication facilities for object to object communication between applications residing in different devices. It supports user-defined application processes. It also supports object-oriented software and predefined data structures for effective, robust, and manageable communication between devices.

The application layer must have an efficient messaging system. It would thus result in faster configuration uploads and would reduce traffic congestion. It thus leads to a very effective and energy efficient communication resulting in enhanced battery life.

26.15.1 STRUCTURE

The application layer of ISA100.11a is subdivided into upper application layer (UAL) and application sublayer (ASL). UAL contains UAPs that address and perform user specific problems. It also supports computational functions, protocol tunneling, input/output hardware needs, etc. Object oriented communication and routing between peer objects belonging to the same or different UAPs across the network is carried out by ASL.

26.15.2 PROTOCOL DATA UNIT

The application layer takes the responsibility of addressing a particular object belonging to a particular application process. Header of APDU contains the object identifier information for addressing the object, identification of service type for specific service needed to be provided, etc. The object identifier addressing scheme becomes very efficient by previously saving the coding bits of the identifiers of both the source and the destinations.

When APDUs need to be concatenated, inferred addressing mode is used. In this, a reference is made between the source and destination object identifiers of APDUs with the most recently transmitted source and destination object identifiers of the APDUs in the same concatenation. This saves the overhead bits of the headers of the APDUs.

26.15.3 COMMUNICATION MODEL

It has three different formats. They are: unidirectional buffered communication, queued unidirectional communication, and queued bidirectional communication.

Unidirectional buffered communication is used by publish/subscribe service for unconfirmed scheduled periodical and aperiodical data publishing

Queued unidirectional communication is used by source/sink service for unconfirmed, unscheduled alert reporting, but without any flow or rate control. But an alert acknowledgement is sent back to the alert source. If no acknowledgement is received, the alert reporting is repeated.

Queued bidirectional communication is used by client/server service for bidirectional communication in cases of on-demand one-to-one aperiodic communication having retries and flow control facilities.

26.15.4 OBJECTS, THEIR ADDRESSING, AND MERITS

Application layer defines standard objects, standard attributes, interoperability issues. Industry and vendor-specific standard objects may also be included in the same layer. These objects are contained in UAPs. A user application process management object (UAPMO) is a must in a UAP. The UAPMO contains number of objects, UAP status, etc. UAP would also contain an upload/download object (UDO) if the UAP has a self-upgrada-tion feature.

Since application layer has many objects, addressing individual objects is a must for proper access and identification of an object. For this, each object has a 16-bit object identifier ID. By default, UAPMO is assigned a reserved ID as 1 in every UAP and an ID of 2 is reserved when UDO is implemented.

Objects residing in the UAPs at application layer have many advantages like: interoperability with different field devices and also legacy systems, modular approach enables add-ons for either industry or vendor-specific objects within the same UAP, better network management, etc.

The user application layer consisting of several UAPs is shown in Figure 26.9. Each UAP consists of several objects and with each object hav-ing its own attributes, methods, and alerts. The application layer is modular, object-oriented, extensible and can accommodate a variety of applications. Multiple applications are supported through UDP port addressing.

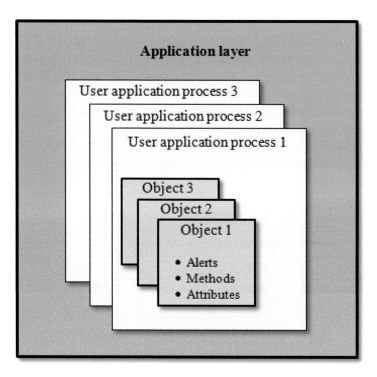

FIGURE 26.9 The application layer. (Courtesy: ISA, The Technology behind the ISA100.11a Standard-An Exploration. ISA100 Wireless Compliance Institute, 2010.)

26.15.5 Gateway

To support gateway functionalities, a gateway has a specific UAP. A gateway is required to interface host-level applications to the wireless field devices or else to the wired field devices via an adapter, which acts as a protocol translator. The adapter is used for converting from wireless protocol to wired protocol and vice versa.

26.15.5.1 Gateway Service Access Point

A gateway service access point (GSAP), also called a gateway high side service interface, is a mandatory requirement which provides the requisite functionalities for connecting the field devices to host-level applications.

The foreign protocol translator, residing on top of GSAP, helps in converting foreign host-level protocols like tunneling to communicate with wireless field devices. Object-to-object communication is provided by the ASL services with the help of GSAP and a gateway application process.

26.16 KEYS IN ISA100.11A

ISA100.11a supports both symmetric and asymmetric key based join process. The join process is governed by administrator of the network which helps in providing the new devices with enough information to join and participate in the network. Once joining is over, the new devices gain access to network communication resources and get the relevant bandwidth which ensures normal operations in the network activity. There are different types of keys: data link key is used for message authentication, master key is used between security manager and device while session key is used for normal data transmissions.

26.16.1 Joining by Symmetric Key—A Comparison Between ISA100.11a and Whart

Both ISA11.11a and WHART support symmetric key based joining. Table 26.3. shows a comparison in the symmetric key based join process between the two wireless protocols.

26.16.1.1 Protection of Join Messages

At the transport level, a UDP checksum replaces the session level security. In ISA 100.11a, there is no session key during the time of joining of a device, the security sublayer does not process any outgoing join request from a device and any outgoing join response from the advertising router. The security header size thus becomes zero. The join key in ISA 100.11a

TABLE 26.3
Symmetric Key Based Joining—A Comparison between ISA100.11a and WHART

ISA100.11a	WHART
The device desirous to join the network must be provisioned with necessary information like join key, network information, etc.	Identical to ISA100.11a
The device acquires advertisements from existing routers in the network. The new device develops the join request from the joining advertisement that contains the configuration specifications.	Identical to ISA100.11a
The join request from the new device is forwarded by the advertising router to the system manager. It (join request) contains both security and non-security information.	The join request from the new device is forwarded by the advertising router to the network manager. It (join request) contains both security and non-security information.
The system manager along with security manager processes the join request and checks of whether the new device has enough security features to join the network.	The network manager along with security manager processes the join request and checks of whether the new device has enough security features to join the network.
After the join request is approved, the system manager responds with a join request to the new device. It is then formally inducted in the network.	After the join request is approved, the network manager responds with a join request to the new device. It is then formally inducted in the network.
For hop-to-hop security, a global key is used for data integrity at DL level before obtaining the DL key from system manager. This DL key is utilized for subsequent secure communications along with the new keys from system manager.	For hop-to-hop security, a well-known key is used for data integrity at DL level before obtaining the network key from network manager. This network key is utilized for subsequent secure communications along with the new keys from network manager.
Join messages separate out the security and non-security information in the different services. Security manager handles the security information and system manager handles the non-security information.	Not so in WHART.
Session-level security is carried out with UDP checksum (integrity) at the transport level.	Join messages are protected by join key. This is temporarily used as session key to ensure end-to-end security (encryption and authentication) required at the transport level.
During the joining process, a challenge-response is exchanged between the new device and system/security manager to validate mutual authentication. This ensures whether the two parties are alive and agree on the same shared key.	Not available in WHART.
It supports asymmetric key-based joining.	Does not have this facility.

provides security cover to the application layer data and other key distribution schemes.

26.16.1.2   Key Agreement and Distribution

ISA 100.11a generates a secret key generation (SKG) which is used as a master key between a device and the system manager. The master key is then used to derive DL key and session key in the encrypted message from system/security manager via *Security_Sym_Join.Response()*.

Sequential steps are needed for key distribution cum agreement scheme required in a symmetric join process. This is realized by SKG methods involving join key.

The device first sends its EUI-64 to the system/security manager via *Security_Sym_Join.Request()* and the 128-bit AES. The SKG generated by the security manager is sent back to the device via *Security_Sym_Join. Response*. The security manager also sends its own 128-BIT AES. After the join key authenticates the MIC of other related data, the master key encrypts session and DL keys. The device shares the same master key as that of the security manager. This is confirmed by the device vide *Security_ Sym_Confirm* with both the device and the security manager sharing the same join key.

26.16.2   ASYMMETRIC KEYS

Unlike WHART, ISA 100.11a supports asymmetric keys, although it is optional in nature. Different keys are used to encrypt and decrypt a message. Each device has two asymmetric keys: a public key and a private key. The public key is openly circulated while the private key, as the name suggests, is kept a secret. A message is encrypted with the public key and decrypted by the secret key. There are two asymmetric keys in ISA 100.11a: CA_root and Cert-A. The former is a public key certified by some authority authorized to certify the device's asymmetric key. Cert-A is the asymmetric key certificate for device A and is used to authenticate the asymmetric key establishment protocol.

26.16.2.1   Security Policy

ISA100.11a has different levels of security for data authentication and encryption at its data link and network layers. The security manager controls policies related to cryptographic material that it generates.

26.17   PROVISIONING OVERVIEW

A new device, while joining the network, must first be provisioned before its formal induction into the network. This device is called the *device to be provisioned* (DBP). The wireless sensor network which the device intends to join is called the *target network*. A device which is able to provision a new device into the network so that the latter can technically become part of the network is called the *provisioning device* (PD). The system/security manager and a handheld device can act as a provisioning device. The PD would implement the DPSO which would provide information to the *Device Provisioning Object* (DPO) in the DBPs.

The network between the PD and the DBP is a Type A field medium for provisioning device OTA. It can be a temporary, isolated small network, or else a separate logical network which works along with the target network consisting of system manager or a security manager. The logical network can have different forms depending on priority and security levels.

The provisioning procedures undertaken at the site may instead be done at the factory premises if the device manufacturer is delegated with such provisioning powers. In that case the device can straightway join the network with the pre-installed symmetric K_Join/PKI (Public Key Infrastructure) certificate and other network related information. It should be noted that the system/security manager is synchronized with the same secret keys.

26.17.1 DIFFERENT KEYS

A secret AES symmetric key is used for data confidentiality which can either be pre-installed at the factory or can be provisioned during network initialization. The system/security manager of the network only can have the knowledge of this key.

K_Global is an asymmetric key based OTA provisioning used for authentication of device credentials and reading of device identity and configuration settings. This key is used for data integrity during provisioning process. In case the target network does not have asymmetric cryptographic support, the DBP joins the target network with the help of PKI certificates and the K_Join key.

K_Join is an open join key and is set to a non-secret value by default. This is used for joining the network provisioning in an unsecured manner.

26.17.2 CONFIGURATION BITS

Four configuration bits A1 to A4 are all preset to 1 by default and have their own functionality. They are as follows:

A1: It allows OOB (Out of Band) provisioning.
A2: It allows asymmetric key based provisioning.

A3: It allows default joining (joining a default network with default join key).

A4: It allows reset to factory defaults.

The default settings of a device can be changed during pre-provisioning stage.

26.17.3   REQUIREMENTS FOR JOINING

For a device to be able to join a network, both network-related and trust-related information must be provisioned into the device. The network-related information required are: Network ID, 128-bit address of system manager, subset of frequencies within which the device is able to listen and respond to correct frequencies, join time, DL configuration settings. Again trust-related information required are: EUI-64 of security manager, specific Key_Join to join a specific network and types of network join methods supported. The process of joining and leaving a network is shown in Figure 26.10.

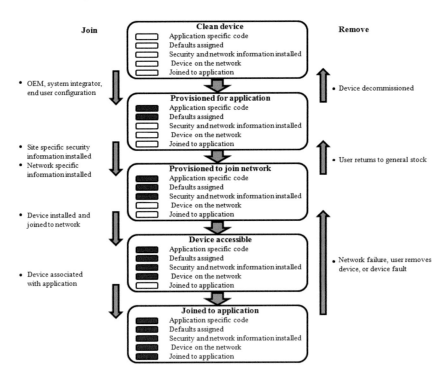

FIGURE 26.10   Process of joining and leaving the network. (Courtesy: W. W. Manges. ISA100.11a Principles of Operation Overview, ISA Expo 2007, Oak Ridge National Laboratory, 2007.)

Several steps are necessary before a device can formally join a network. Application-specific codes and defaults are first assigned, followed by installations of security and network information. Then the device is installed and it joins the network.

26.18   DATA DELIVERY RELIABILITY

Path diversity and duocast are implemented to increase data delivery reliability. Path redundancy in the form of redundant routes is continuously followed to adapt to existing spectrum conditions. It addresses coexistence issues without user intervention. Also the devices make adaptive routing to enhance reliability of data delivery.

ISA100.11a supports mesh topology based on graph routing. Graphs preconfigure data routes set by system manager and data packets follow inbound and outbound routes to reach the final node, i.e., destination.

In Duocasting, a data packet sent by a device is received by multiple devices within the same time slot. It is normally implemented for devices having direct connection with routers but can also be applied for devices which operate deeper in a subnet. Successful data delivery increases exponentially when duocast (or n-cast) is employed.

26.19   TWO LAYER SECURITY

A two layer security is enforced in ISA100.11a for transmission of data message—for hop-to-hop authentication and encryption, data link layer security is enforced while for end-to-end authentication and encryption, transport layer security is enforced. For hop-to-hop transportation of a data message from one node to another in the same subnet, a MAC (DLL) to MAC transfer takes place. Along the transmission path, the routers authenticate and encrypt/decrypt data packets as they progress to the destination. This is shown in Figure 26.11.

For end-to-end security of data messages, authentication, and encryption/decryption of the same is done by the originating device at the transport layer. Only the destination device has the capability of authenticating and decrypting the received data message at its transport layer.

26.20   COMMUNICATIONS IN ISA100.11A

Communication in IS100.11a can take place in various ways: between two nodes belonging to the same subnet; between two nodes across two subnets; from a node to the control system; from a legacy node to the control system.

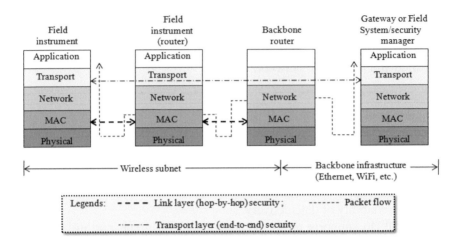

FIGURE 26.11 Two layer security. (Courtesy: ISA, The Technology behind the ISA100.11a Standard-An Exploration. ISA100 Wireless Compliance Institute, 2010.)

The simplest communication process in ISA100.11a is when communication takes place between two nodes belonging to the same subnet. This is shown in Figure 26.12. The message does not have to flow through either the backbone router or the gateway to reach the final destination.

The process of information flow starts at the application layer of the first node, travels through different layers, before reaching the router. It travels through the physical and data link layers of the router before reaching the

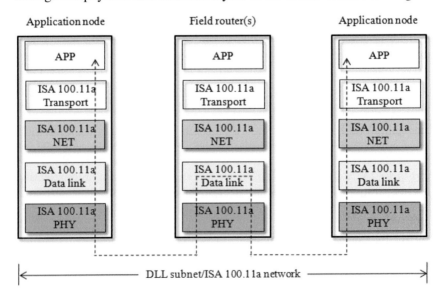

FIGURE 26.12 Communication in the same subnet. (Courtesy: W. W. Manges. ISA100.11a Principles of Operation Overview, ISA Expo 2007, Oak Ridge National Laboratory, 2007.)

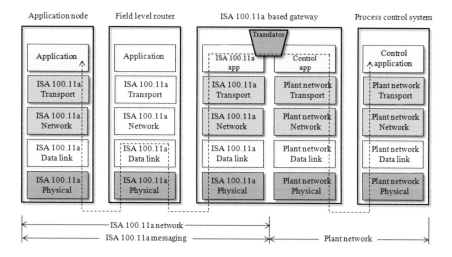

FIGURE 26.13 Communication between a node and the gateway. (Courtesy: W. W. Manges. ISA100.11a Principles of Operation Overview, ISA Expo 2007, Oak Ridge National Laboratory, 2007.)

application layer of the second node after entering it through the physical layer.

Figure 26.13 shows the example when the message from the application node travels all the way to the control system via the field router and the gateway. A translator converts the ISA100.11a application to the control system application and vice versa.

Again, Figure 26.14 shows a data communication scheme which involves data transfer from a legacy device to the control system. It

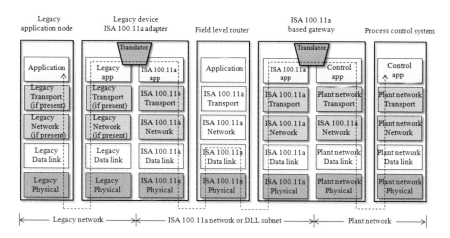

FIGURE 26.14 Communication between a legacy node and the gateway. (Courtesy: W. W. Manges. ISA100.11a Principles of Operation Overview, ISA Expo 2007, Oak Ridge National Laboratory, 2007.)

involves use of two translators—one for translating legacy application to ISA100.11a application and the second for converting ISA100.11a application to control system application. The whole communication process involves three networks: legacy network, ISA100.11a network, and plant network.

26.21 ISA11.11A AND WHART—A COMPARISON

Table 26.4 shows a comprehensive comparison between WHART and ISA100.11a.

TABLE 26.4
Comparison between WHART and ISA100.11a

ISA11.11a	WHART
A mesh, star or mesh-star network	A mesh network
Field devices can act as end nodes with no routing capability and/or as router nodes with routing capability	Field devices and adapters can act as routers capable of forwarding packets to and from other devices
Wireless provisioning	Wired provisioning
Not all devices in the network are necessarily capable of provisioning other devices to join the network	All devices are capable of provisioning other devices to join the network
Application-layer subdivided into upper application layer and application sublayer	Application layer not subdivided
Network layer not subdivided	Network layer subdivided into network layer services and network layer
Data link layer subdivided into upper data link layer, MAC extension and MAC sublayer	Data link layer subdivided into logical link control and MAC sublayer
Operation defined only in 2.4 GHz ISM band. Uses channels 11–25, with channel 26 as optional	Operation defined only in 2.4 GHz ISM band. Uses channels 11–25
Each channel bandwidth is 2 MHz, channels are spaced 5 MHz apart	Each channel bandwidth is 2 MHz, channels are spaced 5 MHz apart
A combination of DSSS along with FHSS is used as modulation technique	A combination of DSSS along with FHSS is used as modulation technique
Data rate: 250 kb/sec, maximum transmitted power 10 mW	Data rate: 250 kb/sec, Maximum transmitted power 10 mW
TDMA with frequency hopping for channel access	TDMA with frequency hopping for channel access
Only one superframe is operational at a time. The same can be added/deleted while the network is operational	Only one superframe is operational at a time. The same can be added/deleted while the network is operational
Time slot value is configurable and set to a specific value by the system manager when a device joins the network	A time slot has a fixed value of 10 ms
It defines five pre-programmed hop patterns	Does not explicitly define the frequency hop pattern

TABLE 26.4 *(Continued)*

ISA11.11a	WHART
Supports counter with cipher block chaining message authentication code (CCM mode) in conjunction with Advanced Encryption Standard (AES-128) block cipher using symmetric keys for message authentication and encryption	Supports counter with cipher block chaining message authentication code (CCM mode) in conjunction with Advanced Encryption Standard (AES-128) block cipher using symmetric keys for message authentication and encryption
Security features are optional	Security features are mandatory
Generation and management of security keys and authentication of new devices are managed by security manager	Generation and management of security keys and authentication of new devices are managed by security manager
Joining by symmetric and asymmetric methods	Joining by symmetric method only

26.22 CONCLUSION

Compared to WHART, ISA100.11a offers a wider coverage for process automation needs. Scalability of the network is derived by dividing the whole network into smaller ones—called subnets which are joined by backbone routers. Subnets minimize latency in the network. Compatibility of ISA100.11a with IPv6 ensures that various internet technologies can better be embraced. Network management of the system is made easier by object orientation and helps improve better interoperability of legacy systems. Better control of field devices is made possible because of the contract between devices and the system manager. Variable timeslots help manage different communication needs.

However, WHART is better positioned compared to ISA100.11a because of huge installed base of WHART compatible devices. It is possible to integrate the two wireless standards by having a dual protocol stack over the MAC layers which would understand messages from different sources and which are in compliance to either of the two standards.

Index